MYTHICAL MONSTERS SUNKEN CONTINENTS AND ALIENS

Martin Thomas

i

Website Address: MartinThomasAuthor.com

Individual copies of these books can be acquired from BarnesandNoble.com

Table of Contents

PREFACE

Millions of people believe that there are mysterious objects flying in the sky. They were originally described as Unidentified Flying Objects – hence UFOs – but have more recently been termed UAPs – Unidentified Anomalous Phenomena. As most humans spend their time on the land, there are many more sightings of UAPs there, than have been associated with maritime UFOs, which have been termed Unidentified Submersible Objects (USOs). The abbreviation UAP is now used to cover both UFOs and USOs.

The term "Alien" has been used to describe any occupants or manufacturers of UAPs. However, this word has many pejorative associations – it means basically "not one of us" and is frequently used to describe immigrants, whose presence may be illegal, or who are disliked because of their race, colour or creed. I intend to use the word "Others", perhaps not an ideal term, but at least it avoids many of the pre-assumptions inherent in traditional terms.

In my previous work[1], I looked at the information available on the sightings of UAPs, and particularly of "Others", to try to understand what they were doing here. Large "Others" were considered sexy, whilst small "Others" were a turn-off. Of course everyone considered they knew everything about those ambiguous creatures the Small Grays, but the word dwarf was used all too often to describe anything smaller than them, so ignoring the major contribution which Dwarfs have made to Human development.[2]

I discovered at least 24 species of "Others" under 1.5m tall, including colonies of Chaneques or Aluxes in Mayan Mexico[3], tiny Elf-like "Others" in Indonesian Java[4], and small Elf-like "Others" in New Brunswick in Canada.[5] I have called this whole group of entities the Fae.

Almost all lists of "Other" species include Reptilians, describing some of these as early inhabitants of Earth, and others as malevolent invaders. My earlier work indicates that some of these are relatively small, coming in at less than 1.5m tall.

It is also clear that lists of "Other" sightings overlapped with Ghost sighting lists. This is an issue which needs to be resolved.

In the 18th and 19th centuries, there were many reports of giant Sea Monsters and Lake Monsters attacking ships, or simply scaring the lives out of sailors. It has been suggested that these monsters could be Extra-terrestrial in origin. How much if this can be the truth? Are they still around today?

Plato may have told us about Atlantis over two thousand years ago, but it appears many others wanted to get in on the act. There are apparently sunken lands everywhere, but what is the truth of it?

Chapter 1
WHAT IS GOING ON?

a) <u>I GET CONVINCED UAPs ARE REAL</u>

Ancient clay tablets[6] found in Sumeria dating back thousands of years suggest that Humans were visited by strangers called the Anunnaki, a race of giants that they worshipped as Gods. When viewed with a modern eye, these would probably be considered as some form of Extra-Terrestrial.

In my first book[7], I reviewed a wide range of UAP sightings to get a feel for their likely veracity. I soon excluded any sightings where there was only one witness, on the basis that I was not able to go into their possible motivations, whether honest, dishonest or delusional. Then, I eliminated mysterious cloud formations and single unmoving points of light, on the basis that they could so easily be miss-identifications.

What remained was a group of sightings which could not be explained away so easily. If it wasn't so worrying, it would be comical: there were desperate attempts by officialdom to account for these, including such fantasies as marsh gas burning[8] high in the air rather than getting diluted as it rose,

1

or the highly improbable chance coincidence[9] of a meteor and an earthquake in a non-earthquake zone.

Then we have threats by mysterious agencies to witnesses and their families, either of violence or of destruction of their careers.

Even if the evidence of these remaining sightings were not so convincing, the desperate attempts at denial would probably have convinced me of the reality of UAPs and "Others".

b) <u>THE FAE</u>

A sizeable proportion of the world believe that, in addition to the physical world in which we live, there is a magical realm which overlaps with ours, so that we can sometimes see the denizens of this realm[10].

I use the English name: The FAE.

These are claimed to vary from Fairies of less than 10cms tall through Gnomes, and Brownies to Dwarfs and Elves nearly 1.5m tall.

I believe in the late Arthur C. Clarke who famously said:

> Any sufficiently advanced technology is indistinguishable from Magic.

I am not like the White Queen in Alice through the Looking Glass who could "believe in six impossible things before breakfast". I can only cope with one at a time.

I remain certain that UAPs and "Others" exist, even if they have been in our folklore as far back as human memory can stretch. I hope that I managed to convince some of my readers in my first book.

In my last book, I looked at the world of the Fae from the point of view of someone who doesn't believe in magic, but does believe in advanced technology. I was able to show that these Fae were really small "Others" using technology to perform feats considered magical. I was able to show there were at least 24 species of smaller "Others" resident on Earth. These are listed in Appendix 1.

In summary, my findings were:

The Fae comprise many species of "Others", which have arrived on earth, probably to set up colonies, at different times over the last few millennia. Their landing places are widely scattered, and they have probably spread to cover those parts of Earth which they didn't reach to begin with. They brought with them their technology which permitted them to travel from their own planet, together with devices to let them fly, disappear, cast illusions perhaps using holography, and to read and project thoughts.

c) LARGER SPECIES OF "OTHERS"

In Appendix 3, I have listed the predominant species of "Others" over the height of 1.5m. These come from a variety of volumes as described in my book "The UFO, ET, Alien Omnibus"[11].

Sometimes I think the names are quite arbitrary – Take any star or creature and make it into an adjective. The KGB list seems only to have Small Grays in common with this list. Some descriptions are flatly contradictory.

With so many "Others" being described as tall Nordics or tall and human-like, you are left wondering how we ever knew that these were different species.

Chapter 2
INTELLIGENT REPTILES

a) VARIOUS CLAIMS

Humanity's shrew-like mammalian precursors survived the mass extinction of the dinosaurs 65 million years ago because we lived in holes in ground. It is easy to believe that some small reptiles survived in just the same way. They could have been more advanced than us back then, or they could have evolved more rapidly than us over those millions of years.

However, there are many alternative theories, stretching from the impossible to the improbable, the unlikely and the un-provable.

If we look first at "The Reptile Ascendancy"[12] by Fufu Fang, this claims that Reptilians have been involved from Human creation right through to the present day, manipulating us and ruling us without our knowledge. This hypothesis appears to ignore all UAP and "Other" sightings around the world. It is mutually exclusive to the "Ancient Alien" hypothesis, to which I subscribe. I therefore consider Fang's hypothesis to be impossible.

5

We next consider the hypothesis put forward by Ana Lorens in her book "Reptilians"[13]. This proposes that Reptiles are capable of shape-shifting and have substituted themselves for all the major Human power-brokers on Earth, and rule us without our knowledge.

Whilst I accept that some reptiles can change colour by choice, it is a big stretch to suggest that they can physically change shape on demand. This would defy some physical laws which I would be unhappy at giving up – for example, if they change size, does their density change or where does the extra mass go to or come from? We know mass can become energy, but where would it be stored? There would be a lot of energy to move around.

They could perhaps use a mixture of image projection and mental control. We know "Others" do this, but it has its limitations. I consider this hypothesis to be improbable.

Then we have the hypothesis put forward by Steve Preston in his book "Lizard People"[14]. Here he suggests that reptiles on Earth are all descended from Lilith, the first wife of Adam (of Adam and Eve fame). She transformed herself into the serpent which tempted them with the fruit from the Tree of Knowledge of Good and Evil. For this, she was condemned by God to remain in this form, as were her reptilian descendents. Preston points out the similarities between fundamentally malevolent deities, around the world, which fall into basically two categories – those he describes as coffee-bean eyed like the Babylonian statuettes, and those which are big-eyed like the Gorgons of Greek mythology. He doesn't really develop his ideas into what may be happening today but, if you are willing to accept as true the story of Adam and Eve, his hypothesis becomes possible.

If you want a Galaxy-spanning tale, then the hypothesis put forward by Maestà Pastore in "Encyclopaedia of Alien Races" is the one to go for. The story here is that some of the many Reptilian races originally occupied Earth until they were driven off in a great war, leaving behind two groups of survivors. One group of bipedal humanoids is hiding underground, waiting for the chance to rebuild their once great civilisation. These are called Saurians. The other group are human in appearance but, having been with the Reptilians for centuries, have the Reptilians' attitude of superiority towards Humans. They are apparently the survivors of Atlantis, and together possess most of Earth's wealth. This hypothesis could even be true, but how do you prove it?

In the world of popular Science Fiction, whenever Reptilians are portrayed, it is almost inevitable that they will turn out to be malevolent entities. This could be racial memory resulting from any of the hypotheses put forward above, so does not help in deciding which is the most probable. Is there any evidence to help?

b) AVAILABLE EVIDENCE

My primary sources of evidence are the books by George Mitrovic which bring together sightings from all over the world of UAPs, "Others" and strange phenomena up to the year 2000. However, great care has to be exercised in using these books because they clearly require the application of far better editorial control. They abound in missing words,

7

typographical errors and duplication of reports, which are sometimes identical apart from dates or locations, and often simply repeated. They are sometimes even in triplicate.

Table 1 – World Sightings of Saurians

Countries/States in the One Volume If Underlined: 2 or More Sightings	Sightings of all sorts	Sightings no UAPs
US West Coast[15] - Alaska, California, Oregon, Washington	5	4
US South[16] – Alabama, Arizona, Arkansas, Colorado, Gulf of Mexico, Kansas, Kentucky, Louisiana, Mississippi, Missouri, Nevada, New Mexico, Oklahoma, Tennessee, Texas, Utah	10	3
US East Coast 1[17] – Connecticut, Delaware, Florida, Georgia, Maine, Maryland, Massachusetts, New Hampshire, New Jersey.	6	3
US East Coast 2[18] –New York, North Carolina, Pennsylvania, Rhode I, South Carolina, Washington DC, Virginia, West Virginia	4	2
US Central 1[19] – Idaho, Illinois, Indiana, Iowa, Michigan	5	2
US Central 2[20] – Minnesota, Montana, Nebraska, North Dakota, Ohio, South Dakota, Wisconsin, Wyoming.	2	2
Hawaii	1	1
US Total	(33)	(17)
Canada[21] – Alberta, British Columbia, Manitoba, NW Territories, Nova Scotia, Nunavut, Ontario, Prince Edward I, Quebec, Saskatchewan, Yukon Territory	7	2
Central America[22] - Bahamas, Belize, Bermuda, Costa Rica, Cuba, Dominican R, El Salvador, Guatemala, Honduras, Mexico, Nicaragua, Panama, Puerto Rico	19	15
[23]Bolivia, Brazil, Columbia, Ecuador, Guyana, Suriname & Venezuela	7	6
[24]Argentine, Chile, Paraguay, Peru & Uruguay	14	10
Russia[25] West	6	3
Central	2	1
East	2	1
Australia[26], *New Zealand & The Pacific Ocean*	7	6
Belarus, Denmark, Finland, Moldovia, Norway, Sweden and Ukraine[27]	8	5

France[28]	4	3
[29]Andorra, Azores, Canary I, Gibraltar, Spain & Portugal	7	5
[30]Albania, Bulgaria, Croatia, Czech R, Greece, Hungary, Italy, Malta, Montenegro, Romania, Serbia, Slovakia and Slovenia	10	8
[31]Austria, Belgium, Estonia, Germany, Holland, Kalingrad, Latvia, Lithuania, Luxembourg, Poland & Switzerland.	4	3
[32]Ireland, Scotland and Wales	1	0
England[33]	6	2
[34]Afghanistan, Armenia, Azerbaijn, Bahrein, Bangladesh, Botswana, Cambodia, Cameroon, China, Congo, Egypt, Eritrea, Ethiopia, Gabon, Georgia, Ghana, Hong Kong, India, Indonesia, Iran, Iraq, Israel, Ivory Coast, Japan, Jordan, Kazakhstan, Kenya, Kuwait, Kyrgyzstan, Lebanon, Libya, Madagascar, Malaysia, Mongolia, Morocco, Mozambique, Myanmar/Burma, Nepal, Nigeria, North Korea, Oman, Pakistan, Palestine, Persian Gulf, Philippines, Reunion I, Saudi Arabia, Senegal, Singapore, Somalia, South Africa, South Korea, Spanish Morocco, Sri Lanka, Sudan, Tajikistan, Tanzania, Thailand, Tunisia, Turkey, Turkmenistan, Uzbekistan, Vietnam, Yemen, Zambia, Zimbabwe	10	7
Totals	147	91

c) ANALYSIS

One way of determining which of the hypotheses in part a) above is the more probable, would be to see whether the supposed Saurians actually exist, hiding away in their underground caverns.

There are two possible sorts of sightings on Earth that could be made by creatures looking like reptiles. They could be the hypothetical ancient Saurians venturing out from their secret underground hideaways, or they could be recently arrived Reptilians either coming directly from space, or from a newly-created Reptilian base on Earth.

9

Of these, the possibly long-established secretive Saurians are the more likely to move around without the use of machinery, confining their surface presence mainly to locations near their caverns, which could be close to current Human centres of population. More recently arrived Reptilians would have been forced to establish their presence away from modern Human centres, and be more likely to need to use UAPs to achieve their objectives, however nefarious or benevolent they may be. Of course, this does not mean that these putative Saurians do not sometimes use their own UAPs.

In the table in Appendix 2, I have listed sightings from all over the world, marking whether their presence is associated with an UAP sighting. In the table below, I have shown the relative percentages of possible reptile sightings in each decade, based on this criterion.

Decade	Pre 50s	50s	60s	70s	80s	90s	SUM
Without UAPs	9	11	11	27	14	19	91
With UAPs	0	3	8	15	15	15	56
Percentage of total with UAPs	0	21	42	36	52	44	

Table 2 Sightings of Reptiles Through the Decades

It is likely that the number of Reptilian sightings, from the seventies onwards, is a big under-estimate as many sightings describe "Others" as wearing either a helmet with an opaque visor, or monk-like garb with a hood. Both of these make it impossible to see their faces. This could be because the various "Other" species, in all their varieties, wanted to avoid scaring witnesses more than necessary, or it could be because Reptilians want to keep their presence secret.

Clearly, on the basis of this analysis, there were few sightings of reptiles associated with UAPs in the early part of the twentieth century, but reptilian UAPs became more prevalent as the decades progressed. Looking at the three hypotheses which I described earlier, that put forward by Maestà Pastore in "Encyclopaedia of Alien Races"[35] seems the most likely. Saurians have been hiding underground for many centuries, and Reptilians have only arrived here in recent times.

d) SAURIAN SITES

If we concentrate on those sightings listed in Appendix 2, and eliminate all sightings where UAPs are involved, we come up with the plots of sightings in Figure 1 for the United States, Figure 2 for Europe, and Figure3 for the Earth as a whole.

The distribution of Saurians around the world suggests that:

Their North America underground hiding places are probably somewhere on the northern West Coast in one of the many karst areas – limestone which has been extensively water eroded[36] - of the Cascade Mountains, and somewhere in the Ozarks of Missouri, and the Ridge Plateau in Pennsylvania. The Florida sighting could come from Cuba.

Their Central America caves are in Mexico, which is one of Earth's major cave regions,

Their South America caves are

in the Argentine with only limited areas of limestone and gypsum, capable of being transformed into karst, but the potential for many caves in the unexplored areas of the Andes,

and Mato Grosso and Bambui in Brazil, where there are areas of karst.

Figure 1 – Saurian sightings in the USA (Total 17)

In Europe, the frequency of sightings around the Alps, particularly in Italy, Spain and France, suggest that they have a refuge somewhere in one of the karst areas of the western Alps. There could also be one in the Pyrenees between Spain and France.

Clearly there has to be some form of Saurian underground retreat in Australia, given its distance from other

12

sightings. There are areas of karst along its south coast, and in Tasmania.

Figure 2 Saurian Sightings in Europe (Total 26)

In the USSR, there would appear to be a cavern in the karst of the Caucasus, and in a similar area just south of the Urals

The various possible underground hiding places, which I have listed above, would account for the distribution of reptile sightings which are not associated with UAP sightings – most likely the reptile species known as Saurians. The other reptile sightings are probably of an "Other" race which has only recently returned to Earth, having been forcibly evicted many centuries ago, as proposed in the story put forward by Maestà Pastore.[37]

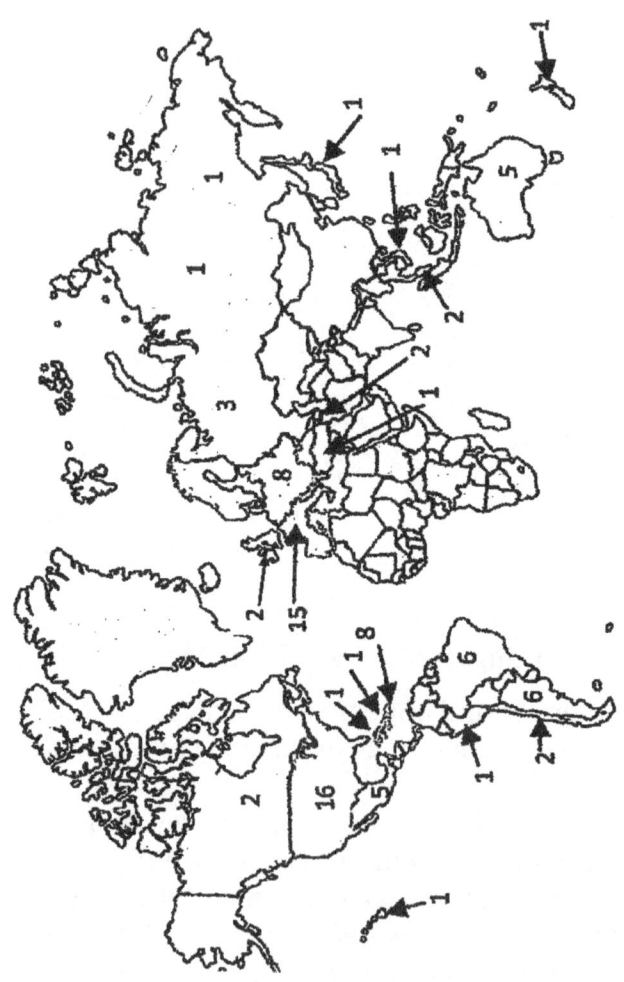

Figure 3 Saurian Sightings Around the World (Total 91)

If we look at a map of the Earth, showing the ice sheets present towards the end of the last ice age (Figure 4), and compare it with the possible Saurian caverns, it seems that they all avoid the ice sheets which were present then. It is possible, therefore that their caverns were carved out during this period. This suggests that the Reptilians were evicted at least 10,000 years ago.

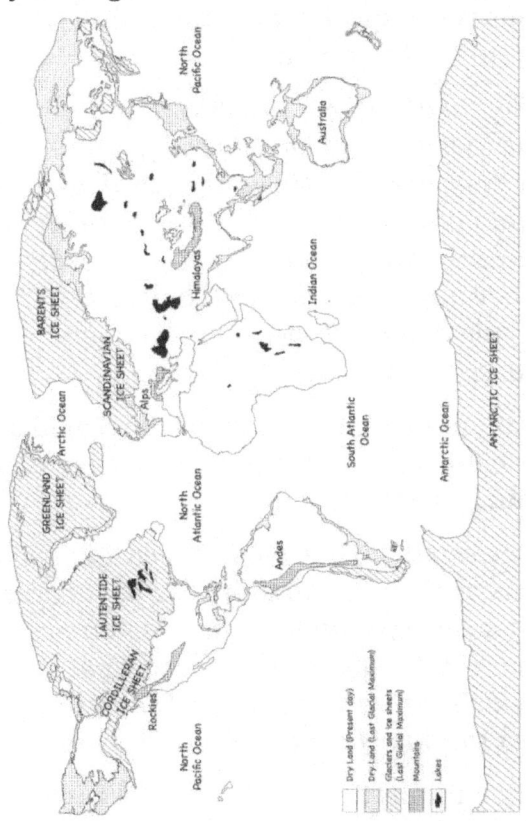

Figure 4 Earth during the last ice age.

15

e) TYPES OF SAURIANS

Looking at the sighting of Saurians, it is possible to draw some conclusions about who are dwelling in these hideaways. It is plain to see that there must be more than one species of reptile falling with the apparently generic term Saurian.

Most obvious is the species sometimes described as having protruding stomachs and buttocks:

> These "frog-like" "Others" are less than 1.0m tall, have large bald heads, with very wide mouths, goggle eyes, and holes for ears. They are sometimes described as having rolls of flesh or fat across their heads. They have an ungainly sideways gait, and appear to have one arm longer than the other. They have webbed hands and feet. In South America they are called Karka[38], perhaps an onomatopoeic tribute to their possible evolutionary origins.

Almost as common are the Saurians described as having a tail. This excludes sighting of say beings like crocodiles where no tail is mentioned:

> These are described as being between 1.2m and 2.0m, scaly with claws or flippers, dark skin, large red eyes, no ears, ridges on its head, a flat nose, and sometimes described as having a crest running from its head right down its back to its tail, which can vary in length. There is enough variation to suggest this may be more than one species.

Many other sightings describe a "huge" lizard over 2.0m in height without going into any further detail.

Interestingly, there are also sightings of lizards with avian facial features, such as a hooked beak. They are sometimes described as having one or two horns on their head.

Clearly, what are described as a race of Saurians are really a community of various types of reptile, who presumably co-exist in the caverns where they sought refuge 10 thousand years ago.

f) WHY ARE THE REPTILIANS HERE NOW?

If the tale from Maestà Pastore[39] is true, the Reptilians were thrown off Earth a long time ago. They appear to have returned to Earth at the first opportunity, and seem reluctant to reveal their presence to Humans. The do seem to be associated with a number of abductions and rapes, co-operating with Small Grays.

Perhaps their ultimate aim is to return to their preferred status quo – where they are once again overlords of earth. To achieve this, they will need to avoid the attention of "Others" who might have different ideas. The principle of non-humans wearing hoods or helmets, in order to avoid frightening us, is working in their favour. Perhaps the Nordics, and Clarions are not even aware of their presence. They tolerate the Small Grays because of their deal with the US[40], and the Reptilians are getting a free ride.

g) SUMMARY

Four suggestions about the source of reptile sighting around the Earth have been proposed. Three are probably impossible to prove or disprove. The fourth could be demonstrable, depending on whether there really are two types of reptile on Earth.

This hypothesis claims that the Reptilians were driven off Earth a long time ago, by "Others" seeking to develop the Human Race, although some Reptilians remained, hiding undetected underground.

I was able to show that there was a significant difference between reptile behaviour before and after the beginning of the nuclear age. Those reptiles in hiding on Earth for many years seldom, if ever, used UAPs. These are the Saurians. Those reptiles using UAPs, who only really showed up in the 1940s, are the Reptilians.

Looking at non-UAP sightings, I have been able to identify possible Saurian hideaway locations around the Earth, and to identify what some of them look like.

However, the Reptilians are back, and are riding on the coat-tails of the Small Greys, probably looking to re-assert their domination of Earth.

Chapter 3
HAVE THEY COME TO HAUNT US?

a)   A SHORT HISTORY OF GHOSTS WORLDWIDE

When the Annunaki left Sumeria about 5,000 BCE, as I describe in my Trilogy[41], they left behind an advancing agrarian society, with a profound belief in magic. This was inevitable, given all the high-tech marvels which Humans had seen them perform, but could not explain.

It is not surprising, therefore, that we can today find clay tablets[42], dated to the 7th century BCE, which describe the practice of Necromancy, how to raise Ghosts and to control them.

At the same time, there arose a belief in Demons, particularly in the Middle East and the Far East. They were not considered particularly evil, but were feared because they did the dirty work like taking reluctant souls into the underworld.

The first recorded Ghost story, dating from about 1,100 BCE, was found on a pottery fragment.

Many cultures believed in an afterlife, and some developed the idea of re-incarnation, but necromancy was mostly condemned by the likes of Judaism, Islam and Christianity. From China and the Far East came the veneration of one's ancestors and respect for one's elders.

Worldwide, there was a general belief in Spirits in nature, protecting trees, streams, mountains etc. In many places, this belief persists to this day. In classical Greece there were so many named Nymphs, they were numbered in their thousands.[43]

b)   GHOSTS OF THE BRITISH ISLES

Certainly every country in Europe and North America would claim that there are hauntings in some buildings and in places outdoors. In some other countries, Spirits frequent places, much as described in ancient Greece with its multitude of types of Nymph. These could well be Elves or similar "Others" trying to protect their environment

I am going to use the British Isles as a fairly well enclosed sample area, where there are relatively densely populated parts such as most of England and the Isle of Man, and some isolated areas in Ireland, Scotland and Wales.

In his book, "British Ghosts and Where to Find Them"[44], the parapsychologist Nick Tyler helpfully categorizes ghosts, firstly into Unintelligent and Intelligent Spirits. The latter are able to communicate with the world around them and seems to understand that they are in the spirit world. Then he sub-divides the former to include Echoes which always appear in the same pattern, and the rest which are not bound by a single pattern of movement.

He then outlines an hypothesis commonly held among parapsychologists - the Stone Tape Theory. In this it is proposed that, just like a old-fashioned tape recorder, inanimate objects, such as stone, might be able to store the residual energy from some events, and could then be triggered to replay, given the right circumstances.

This could account for all the Echo-type hauntings, and perhaps the rest of the Unintelligent-type hauntings, if there were multiple such recordings in the one building. None of these hauntings could be considered to demonstrate the presence of spirits.

What we are the left with are Poltergeists, Intelligent Spirits and Partial-Bodied Apparitions. It is argued that these do demonstrate the presence of a spirit world.

Examples of these are in Tyler's book and two others which are also serving as a source of reference. Listed in Appendix 4 are those instances which cannot be excluded using the Stone Tape Theory, and are not sightings of the 1.1m tall Black Dog known as a Shuck.

In Nick Tyler's book, there were 238 hauntings described. Of these:

12 were of the Shuck. 20 were of activities which could be attributed to an unseen Poltergeist, even though there may have been apparitions of Echoes or Unintelligent Spirits in the same location, but which could not be directly connected to them. 14 were of Intelligent Spirits. These last need to be examined individually

The remaining 192 are of Echoes or Unintelligent Spirits, which can perhaps be dismissed as caused by energies as described in the Stone Tape Theory.

In Sean McLachlan's Book[45], there are more Poltergeists at work than Intelligent Spirits. I have added both of these types of sighting to the list in Appendix 4.

In Ruth Roper Wylde's book[46], which specifically covers sightings on British highways, the vast majority are Echoes or Unintelligent Spirits. This suggests that, although the theory is called the "Stone Tape Theory" it would need to be broader based, covering water, road-building materials and even rocky outcrops.

Overall she reports some 450 sightings, amongst which are 14 Shuck, 5 Poltergeists unusually acting outdoors, several sightings which UAP investigators might claim as theirs, and one possible Intelligent Spirit which could equally count as a real person, because he was never seen to disappear.

The 26 reports of Shuck stand out, because they are the only claimed Ghostly sightings of creatures which we would not have encountered in our everyday lives over the years. UAP investigators also count these, amongst their "Other" sightings, as cryptids which could be from other worlds.

These Shuck are said to be Black Dog-like creatures, walking on all fours, and which are often described as guarding a locality. They are 1.1m tall to the shoulder and it is sometimes claimed that they have partly Human faces, although they are also sometimes mistaken for real dogs, so this cannot always be the case. It is said that they are friendly to the good at heart, and threatening to the not so good.

There is also one report of the infamous "Spring-Heeled Jack", usually reported in lists of "Other" sightings.

Again, I have added the relevant records into Appendix 4.

c) <u>POLTERGEISTS - AN ALTERNATIVE EXPLANATION</u>

In my previous book[47], I stated that I could not behave like the White Queen in "Alice Through the Looking Glass" and "believe in six impossible things before breakfast". I have limited myself to the single belief in UAPs and their occupants. I therefore need to see whether a belief in Ghosts is really necessary.

If I once accept the hypothesis that all Echoes or Unintelligent Spirits can be explained in terms of an enhanced version of the Stone Tape Theory, I need only to account for the remaining sightings of Intelligent Spirits and reports of Poltergeists.

To fit Poltergeists in with the "impossible" idea of UAPs and their occupants, we need to find "Others" who are tricksters, no matter whether amicable or malevolent. Many things may be said about the Small Grays, but they have never been accused of having a sense of humour. Equally the taller "Others", listed in Appendix 3, seem to take themselves very seriously.

 Looking at the relative presence of the Fae and the taller "Others" in the table in Appendix 5, we can see that before the Second World War, the Fae were in the majority in much of the British Isles, with "Others" accounting for only 40% of observations. After Hiroshima and Nagasaki, this rose to 60%, and then fell back to 43%, a pattern echoing that of the reptiles shown in Table 2. However, there was a huge increase in abductions by Small Grays and in UAP sightings. The interests of the tall "Others" are clearly demonstrated, and this does not include playing tricks on Humans.

However, the Fae are another matter. Within the British Isles, amicable species such as Hobgoblins, Pixies and Brownies have been known to play tricks on Humans, particularly if they have been offended and are acting as Boggarts. A Goblin does not even need that excuse and can be quite malevolent.

At one time, almost every home welcomed a House Elf or a House Goblin, but these have been progressively driven out as technology advanced in the home. It has been suggested that they have almost died out, but perhaps not. They could have simply moved on.

It pays to remember that the Fae are not magical beings[48], but technological species with devices giving them capabilities such as that of flying, invisibility, casting illusions and even reading and controlling our thoughts.

Using these devices could well cause some of the effects beloved by parapsychologists, such as detectable Electro-Magnetic signals or localised cooling of the atmosphere. When they wish, they can either make noises themselves, or make witnesses believe they hear them. Outdoor Poltergeist activity could also be the actions of Goblins, who like to aggravate Humans at every available opportunity.

This would suggest that these members of the Fae have moved from humble Human dwellings into the grander buildings where today we get Ghost and Poltergeist sightings. At the expense of the occasional apparently supernatural activity, they are free to live their lives in peace.

d) INTELLIGENT SPIRITS – AN ALTERNATIVE

Going through their 63 hauntings listed in Appendix 4, we can come up with a list of capabilities which an Intelligent Spirit needs to have in the British Isles:

> They must be able to make themselves invisible, or to make only part of themselves visible;
> They should be able to change their appearance to look like someone else;
> They should be able to talk in English, or to project their thoughts so that they appear to be talking;
> They should be able to touch people and to move things when necessary.

This is the specification for Intelligent Spirit Mark 1, and it is immediately apparent that this specification can easily be met by a number of types af Fae, if they used some of their technology. It really only requires them to want to act as a Ghost.

Given what we know about the behaviour of Brownies, Hobgoblins and Goblins, they would enjoy playing the occasional trick on we poor Humans. Boggarts would also enjoy the fun.

I feel that it is almost certain that a Goblin is behind the sighting at Tondu in Wales. He would delight in picking a fight or two.

e) EXTRAPOLATING WORLDWIDE

I see no reason why the findings within the British Isles should not be applicable to the rest of the world. The Fae have been shown to be everywhere[49] and the Stone Tape Theory, if true, would be applicable anywhere on Earth.

f) SUMMARY

I have been able to show that, although Ghosts are claimed to be an Earth-wide reality, in fact it is possible to blame all the sightings either on an effect called "The Stone Tape Theory", or the actions of some "Others".

In the Stone Tape Theory, beloved by parapsychologists, it is proposed that, just like a old-fashioned tape recorder, inanimate objects, such as stone, might be able to store the residual energy from some events, and could then be triggered to replay, given the right circumstances. This could explain any Ghostly sounds or sights where there is no interaction with witnesses.

In any haunting, where either a Poltergeist is at work or the Ghost appears to be aware of the witness, I suggest that it is really a smaller "Other", often known as a Fae. These are known to enjoy taunting or playing tricks on Humans.

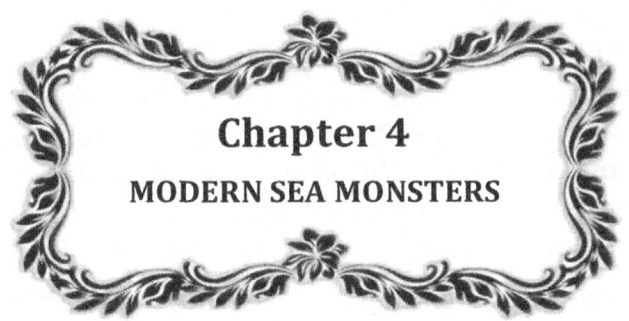

Chapter 4

MODERN SEA MONSTERS

a) <u>THE VARIOUS TYPES</u>

Traditionally, there have been multiple types of sea monster but, in modern times, there have been far fewer sightings. Perhaps this is partly due to changes in fashion, as seafarers do not earn fame and glory by making inflated claims of their bravery in venturing into the unknown. Indeed present day deep sea sailors risk ridicule by reporting what they believe to be genuine sightings.

Nevertheless, there have been sightings reported in the twentieth century which deserve investigation. These are:

The Sea Serpent
The Modern Dinosaur
The Giant Octopus
The Colossal Squid
The Globster
The Giant Turtle
The Whale

The question arises as to whether these monsters are the products of evolution on Earth, or are "Other" species which have taken up residence here, using underwater portals to appear and disappear. This is one of the favourite suggestions to account for the difficulty in finding the Loch Ness Monster – Nessie.

As with UAPs, the sightings are generally so brief that witnesses are taken unawares, and photographic evidence of sea monsters is either totally absent or of inadequate quality to convince sceptics.

One problem is that some of these monsters appear similar to types of dinosaur living before the great extinction event 60 million years ago. However, the claim for the total extinction of sea dinosaurs is not as convincing as that for land dinosaurs. We know that at least one survived. There was a giant turtle, *Archelon Ischyros,* in the late Cretaceous, which was 4m across and weighed in at an estimated 3.3 tonnes. We still have large turtles, such as the leatherback, today. There were snakes living at the same time as the dinosaurs, and they are still with us today.

b) SEA SERPENTS

There have been claims of sightings of large Sea Serpents well into the twentieth century, even though there are people who consider all such sightings to be mistaken identity.

In his book "A Natural History of Sea Serpents"[50], Adrian Shine does his best to dismiss every claimed sighting as deceitful, delusional or mistaken identity, so maligning those possibly credible witnesses who have made sincere reports. I'll come back to his comments about possible modern dinosaur sightings later.

What is known is that, before and after the extinction of the dinosaur, sea snakes flourished, with one – *Titanoboa*[51] – living about 50 million years ago, reported to have grown to 13m in length. What if these monsters didn't become extinct then, but evolved further, and grew bigger. After all, we were only small shrew-like mammals then, and we've changed a bit.

There is one deep-water creature, the Oarfish, the longest boned fish known, coming in at a claimed 15m, with stranded relics being measured at 10.5m. When on the surface it tends to swim with its head out of the water, and that could be taken for a horse's head.

There have been many sightings of Sea Serpents in the early 20th century, but fewer more recently, so perhaps they have learned to avoid us. A list of such sightings is given in Appendix 6.

In his book, page 141, Shine quotes Dr Leslie Noe of the Cambridge Sedgewick Museum[52] as saying that the articulation of the neck bones in *Muraenosaurus*, a type of plesiosaur, could not lift its head up swan-like out of the water, so ruling out the reptile as a candidate for the Loch Ness Monster. I hope that Dr Noe was aware that this dinosaur was alive in the mid-Jurassic, mid-Callovian period, some 160 million years BCE, and hence a100 million years before the dinosaur extinction. By the time of the extinction, *Elasmosaurids* and *Leptocledians* were the dominant descendants of the plesiosaur.[53]

I would be loath to make such a categorical dismissal of Nessie given that, assuming that some plesiosaurs lived through the extinction event, as Turtles and Sea Snakes did, they would have had 60 million years to evolve this capability, should it have proved advantageous for their survival.

If this is the level of scientific scrutiny Shine applied in his determination to disprove the existence of Sea Serpents, I propose to take his conclusions with more than a pinch of salt.

c)   THE MODERN DINOSAUR

Clearly, there is the possibility that some form of plesiosaur, a sea-dwelling creature, could have survived the extinction event, and 60 million years is a mere instant in the life of such species. The question is really how could they have survived in Loch Ness for that long?

Figure 5 A modern dinosaur

The simple answer is that they didn't. In Appendix 6, I list sightings of creatures like Nessie, world-wide. They exist everywhere. There have been sightings of Nessie throughout Scotland's Great Glen, and in the river Ness which links Loch Ness to the sea. It is possible that this Loch, and other similar features around the world, are used as nurseries. There have been sightings of smaller versions accompanying full-size Nessies in Loch Ness, and of smaller versions sunbathing on river banks in Venezuela. Were these dwarf varieties or youngsters?

One anomaly exists which needs to be explained, is that Nessie is sometimes described as having stubby horns. Are these really horns, or perhaps tufts of some sort? Either way, could this be a matter of gender difference?

It certainly seems likely that we have a world-wide breeding population of Nessies, who have also learnt to keep out of the way of Humans.

d) SQUIDS AND OCTOPUSES

i. Ancient History

A cephalopod is any member of a class including squid, octopus, cuttlefish, or nautilus. They are widely regarded as the most intelligent of the invertebrates and have well-developed senses and large brains. The nervous system of cephalopods is the most complex of the invertebrates.

The sequencing of a full cephalopod genome has remained challenging to researchers due to the length and repetition of their DNA[54]. The genome showed similar patterns to other marine invertebrates with significant additions assumed to be unique to cephalopods. Cephalopod genomes are generally large, with some, like the squid, being larger than the Human genome. Studies have revealed large-scale genomic rearrangements, which likely contribute to cephalopod morphological innovations.

Research has shown that Octopuses can form concepts, plan for the future, generate a cognitive map, self-monitor for apparent pain and manipulate communication.[55]

Scientists have had great difficulty in fitting these creatures into any family tree, and it has been suggested that this is because they did not originate on Earth, although they may have been here for millions of years. Of course, there is another possibility. Someone in the past could have been playing about with their genes!

Whatever the reason, they are here now, are extremely intelligent and have the ability to manipulate their environment. Not needing to breathe air, they could dwell deep underwater and there could be species which we have never encountered.

ii. Octopuses

Octopuses have eight arms with suckers on them to grasp their prey. They have large eyes, and no skeleton. They have the ability to camouflage themselves with a chameleon's ability, but much more quickly. They swim using a siphon tube to squirt out water, jet propelling themselves.

Figure 6 An Octopus

In the words of Erica Edwards, in her book "Octopuses"[56], their nervous system is unlike that of any other animal on earth, with a decentralized brain. Two-thirds of its neurons are located in its arms, making them at least semi-autonomous. They also have a central brain, and optic lobes behind their eyes.

They are generally thought to be related to molluscs, but they and their other Coleoidea relatives seem very distant relatives. There are two sub-species, the deep-water ones and the shallow-water ones, which include the largest known Octopus – *Enteroctopus Dofleini*[57] or Pacific Octopus, weighing up to 50 kg, with an arm span of over 5m.

Amongst the sightings of Globsters is one discovered on Anastasia Island off the Atlantic coast of Florida in 1896.[58] It was claimed that this was a giant octopus weighing 5-6 tonnes. This claim is still the subject of heated dispute, although no-one can make a definitive identification.

Octopuses are remarkable creatures, but the known varieties do not, in any way, measure up to their close relations, the Squid.

iii. The Squid

Squid are related to Octopuses, but are much larger. They have eight arms too, but they also have two longer tentacles to catch their prey, instead of using their arms.

The largest squid known in the twentieth century was the Giant Squid, *Architeuthis Dux*, which weighed in at typically 275kg, and about 13m long, thanks to its long arms. In his book, Richard Ellis[59], a noted writer on marine subjects, expressed his doubts about whether there could be anything bigger.

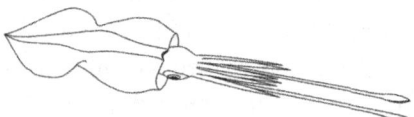

Figure 7 A Colossal Squid

I doubt he was too upset when proved incorrect, on the confirmation of the existence of the Colossal Squid, *Mesonychoteuthis Hamiltoni*, first as bits of tentacle or mantle, and then as a complete specimen. The largest confirmed Colossal Squid specimen to date weighted twice as much as the largest Giant Squid at 495kg, and this is thought to have been a juvenile[60].

The Colossal Squid is not as long as a Giant Squid, but it makes up for this with a huge body and mantle. It is thought a healthy adult might weigh as much as 700kg – 800kg. For comparison, the largest land predator the Polar Bear, weighs only 450kg – 600kg. This squid's body, or mantle, was 4.2m long, about the same as the largest Giant Squid mantles observed, even though this Colossal Squid was a juvenile.

The Colossal Squid lives in the ocean depths, mainly in the Antarctic Ocean where there is sufficient food to supply them. It is described as the ultimate predator, with eyeballs the size of basketballs, so that it can hunt in almost total darkness. Whilst its arms may have conventional suckers, the two tentacles have multiple hooks, to stop its prey getting away.

The best estimate for a Colossal Squid's life span is an astonishingly brief 3 – 5 years, during which time it grows from a millimetre-sized hatchling to a 500kg adult! Its food supply must be pretty good. It is thought that they then mate and die.

Their one main predator which we know of is the Sperm Whale, which dives down into the deeps to hunt Giant Squid and Colossal Squid. We know of this, because Sperm Whales have been observed with rows of tentacle hooks in their skin, where their food has fought back. Also, Colossal Squid beaks have been found aplenty in Sperm Whale stomachs, some much bigger than those found elsewhere. Perhaps there are still bigger Squids to find.

iv. <u>Globsters</u>

These decaying mounds of flesh have been found around the world, and have been sometimes used to argue for the existence of even larger undersea monsters.

It would appear that this is a generic term for any unidentified pile of flesh found on the seashore. In his book[61], Michael Newton list 132 globster sightings worldwide, many of which are vertebrates, and hence not cephalopods. It is not easy to accurately identify the origin of a decaying lump without flesh samples suitable for DNA analysis.

The description of one found in New Zealand is: *a boneless 10m body, 3m high with hairy skin 6mm thick, followed by a layer of fat then solid meat.*

This could easily be the carcass of a Colossal Squid or a Whale. There may be larger undersea monsters out there, but these are probably not their carcasses.

v. <u>Kraken</u>

Despite some claims from the Octopus lobby, there are no known possibilities to explain its existence apart from the Colossal Squid or the Giant Squid. Despite the Kraken's endurance as a seafaring myth, and wonderful illustrations of it attacking and sinking ships, there is no evidence to support its presence beyond these[62]. That doesn't prove that they can't exist, though.

e) <u>TURTLES</u>

As I have already said, it is accepted by palaeontologists that Turtles survived the dinosaur extinction[63], and are around to this day. I list a few sightings in Appendix 8.

One in New Zealand in 1883 was reported to be so large at 20m x 12m that, if this is correct, it could be mistaken for a submarine, or perhaps an underwater UAP.

Figure 8 A Giant Turtle

Perhaps there is more to be discovered about giant turtles but, given that they are air breathers, and their smaller relatives lay eggs on sandy beaches, there really ought to be more recent sightings.

They would probably need to have evolved so that they did not need to lay their eggs on the seashore. A more difficult activity for air-breathers, but maybe they have evolved gills too.

f) <u>WHALES</u>

In his book "The Naked Ape[64]", Desmond Morris suggested that, at one stage, early in their development, proto-humans probably dwelt on shorelines, foraging for fish, shellfish, seaweed etc. To offer proof, he points out that what hair we and some water mammals have on our bodies is very different from other animals. It is streamlined from head to toe, perhaps to offer less resistance when swimming.

If proto-humans were one-time beachcombers, it is not beyond the realms of possibility that two strains diverged here, leaving no fossil record. Some strains opted to remain in the water, remaining oxygen-breathing mammals, and evolving into dolphins, porpoises and whales.

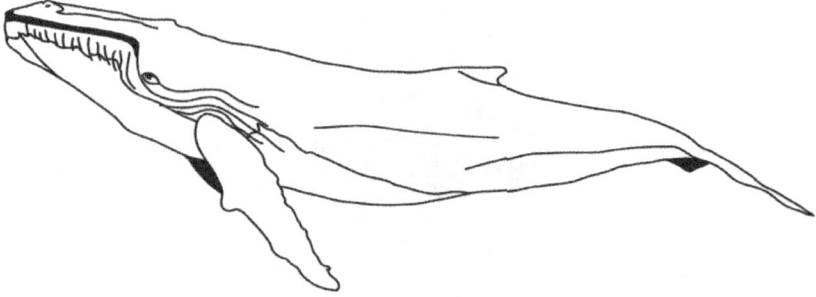

Figure 9 A Blue Whale

We know that some species of cetaceans are very intelligent, and one can speculate that one deep-diving group might have become far brighter than us. Some whales can dive extremely deep, and stay under for long periods. They may have managed to avoid us entirely.

In the 1950s, John Lilly's[65] research helped to advance the view of dolphins as intelligent and socially aware creatures, a perspective that was a marked contrast to earlier, negative views of them. They have been shown to be quick learners and innovative, capable of teaching skills to others.

Even if they do not possess hands to let them become technologically developed, who can say how far their mental abilities could have advanced?

The one thing of which we are certain is that the Blue Whale, checking in at up to 200 tonnes, is the largest creature ever to have lived on Earth[66] – a true monster! It filter-feeds on tiny creatures called krill. One of its smaller relatives, the Sperm Whale, relatively tiny at only 50 tonnes, feeds on giant and colossal squid by diving down to the bottom of the ocean, where they engage in battles which sometimes leave sucker marks on their skin. We will probably never know if they ever lose any of these battles.

g) SUMMARY

Some sea monsters such as Turtles, Snakes, Octopuses and Squid are known to have survived the dinosaur extinction event. The massive Blue Whale has evolved since that time, and is the largest mammal known to have ever existed.

If some survived, why not others? They have also had plenty of time to evolve since then, so they might not be quite what we expect.

Certainly there are a huge number of sightings of a Modern Dinosaur similar to the Plesiosaur, and sufficient sightings of Sea Serpents and Giant Turtles to make one wonder.

The discovery of the Giant Squid and the Colossal Squid has surprised many, proving that evolution could have been working its magic. We don't really need Globsters or Kraken to amaze us, but the possibility is always there.

One thing is for certain however, these sea monsters are all of the home-grown variety. They are not Extra-terrestrial in origin, they are Intra-terrestrial.

Chapter 5
Lost Continents

a) HOW MANY CONTINENS ARE WE MISSING?

There are three principal lost continents: Lemuria, Mu and Atlantis. There are many other areas reputed to have disappeared in ancient times, such as Cantre'r Gwaelod[67] in Wales, but many, if not all, of these are on areas of coastline which could have been flooded when the sea-level rose at the end of the Ice Age. In the case of Cantre'r Gwaelod, sunken forests can still be seen at low tide.

b) LEMURIA

a) Introduction

There is confusion about the location of Lemuria. Many authors are happy to place it in the Pacific Ocean, apparently along with Mu, but it is on record that its presence and its name were first suggested in 1864 by a zoologist, Phillip Sclater[68], suggesting that there had to have been a land bridge between Madagascar and Pakistan to account for the presence of early Lemur bones of *Bugtilemur* dated back 30 million years, and similar to *Cheirogaleus* (Mouse Lemur) bones in modern day Madagascar. He called this putative land mass *Lemuria* for obvious reasons.

This was before the concepts of Tectonic Plates and Continental Drift[69] were developed and put forward as an alternative explanation.

The name was apparently hijacked by a Madame Blavatsky, founder of the Theosophical Society, who believed that it was the home of the root race of Humans, a highly evolved race that possessed psychic abilities, and who were highly skilled in the use of crystals for healing, creating portals and other esoteric practices.[70]

She wasn't particularly concerned about its location, as it was more a idealization like Paradise, or Shangri La. But from there on, it was convenient for writers to attach every available legend to the name Lemuria:

In ancient Hindu texts, there is an idyllic land called Kumarikandam[71], which sank beneath the waves. Obviously this is Lemuria!

The myths surrounding Mu, as proposed by James Churchward[72] are clearly better attached to Lemuria, even though it is believed to be in the Pacific Ocean!

In the Americas, the Hopi Tribe has verbal history of the Lost Continent of Mu[73], which obviously refers to Lemuria, even though they are two oceans apart!

And so on.

Many books are published today with Lemuria and the Pacific Ocean linked together from the title onwards, as the authors seem unable to differentiate between the concept and the reality of the word.

b) <u>The Romantic View</u>

In her book on Lemuria[74] the author pays only lip-service to the possibility that it may be no more than a string of islands. In practice, with very few exceptions, every time she cites some form of physical evidence, she doesn't say where it is located, she simply says "in Lemuria".

About the only places she mentions are Madagascar and Java, which are so far apart that I am left wondering whether she thinks Lemuria was the size of the Indian Ocean. We do know that, during the last ice age, a lot of Indonesia and the Philippines were one land mass. In the same way as some Pacific Ocean enthusiasts claim Lemuria is there, she is claiming Mu for the Indian Ocean.

Two instances of the "evidence" she cites are:

There are petroglyphs with similar designs in many places in Lemuria, proving a unified culture – there are many similar petroglyphs in Greenland and the Azores. Just how big is Lemuria?

There were examples of craftsmanship in bronze, wood and stone, together with examples of polished crystal shards – so they were still in the Bronze Age.

In general, she would quote some fact, not in the least connected with Lemuria, say it was, and then zoom off into flights of fancy, building her castles in the air.

She dates the start of Lemuria to about 50,000 BCE, well within the accepted dates for the Stone Age in the rest of the world, and when hunting and fishing had long been Human practice. However, the earliest archaeological evidence for Human presence on Madagascar is some chopped and scraped bones dated 10,000 BCE[75] – the period when sea levels were rising at the end of the Ice Age. Ironically, this is the date used by the author for the end of the civilization of Lemuria. The best that can be said is that maybe these early Humans were fleeing drowning Lemuria.

The author has attempted to provide a solid foundation for the Lemuria myth, but has failed to do this. The myth is important to some Humans, but the need for its possible reality is less so.

c) Could a Physical Lemuria Have Existed?

160 million years ago, the huge conglomerate land mass, Gondwana, started to split up and then, 135 million years ago, Madagascar separated from Africa but stayed attached to India. About 88 million years ago, Madagascar separated from India, and settled into its current position near to Africa.

India continued moving north until it collided with Eurasia, but there is no general agreement about when this occurred. Estimates vary between 50 and 25 million years ago.[76]

This timeframe will have a major impact on where certain species could be found at any time. It is generally accepted that lemurs evolved in Africa and crossed to Madagascar, swept there on rafts of matted vegetation[77] 60-50 million years ago. There were no lemurs on Madagascar before then. It is thought they were the first primates to arrive,

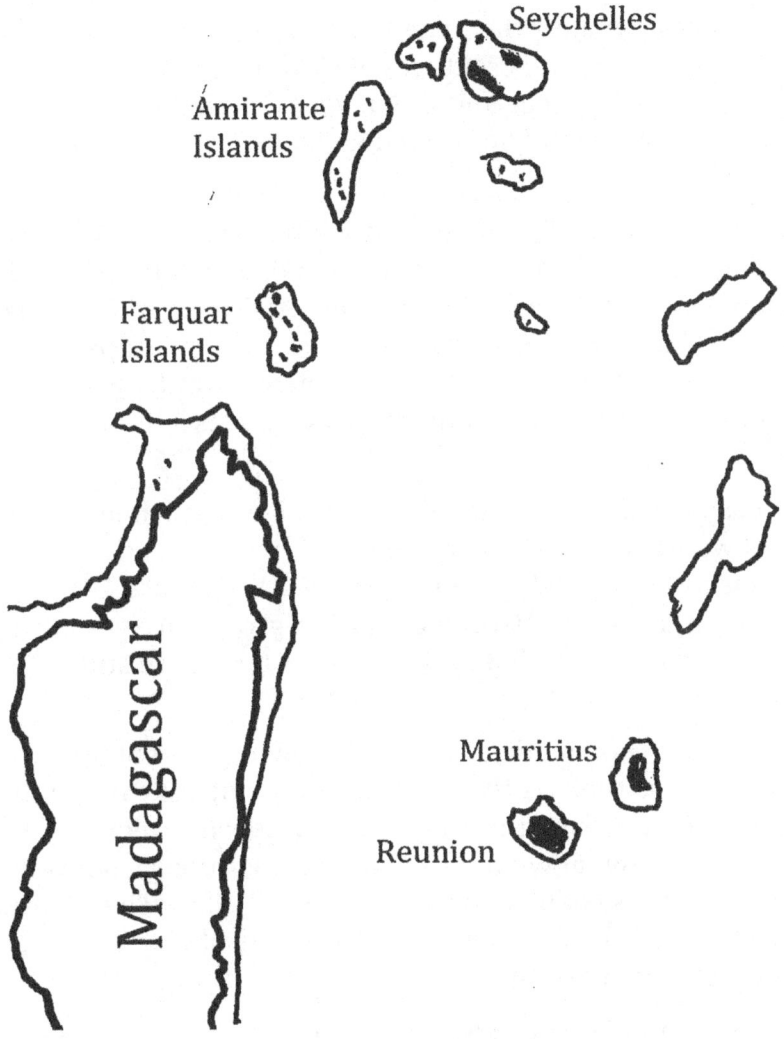

Figure 10 Madagascar, The Seychelles & Mauritius

relatively soon after the mass extinction which caused the demise of many land creatures, including the dinosaurs. By this time, Madagascar and India were distant from each other, so there could have been no lemurs on India then either.

This means that a lemur could not have travelled easily from Madagascar to India during this period. Any lemur in India could only have travelled there from Africa via the Eurasia plate, unless there was either some form of, perhaps intermittent, land-bridge between them, or India was trailing a land-bridge behind it, whilst it was still moving northwards.

A possible solution is that, during the ice-age, when the sea level was perhaps 120 metres lower[78], there was a partial land-bridge as far as the Seychelles Islands (See Figure 10), although the gaps between the island groups are very deep. The Seychelles, up till 30 million years ago, were still attached to the Indian tectonic plate[79].

Lemurs were on Madagascar and evolving from 50 million years ago, so there is the chance they might have got to the Seychelles, crossing the stretches of water between Madagascar, what we now know as the Farquar Islands and the Amirantas Islands, to reach the Seychelles. These straits are today much narrower than the Straits of Madagascar, although much deeper.

However, the presence of these two island groups suggests that, even after the Indian plate and Madagascar split apart, there was still a trailing land link between the two, joining the north of Madagascar to the Seychelles. Certainly the Indian plate increased its northward speed[80] from about

80mm per year to over 180mm per year following the split between it and the Seychelles.

In the interval between their arrival on Madagascar and the time of this split, early Lemurs would have had a chance to cross this link. This route would have brought them to the Seychelles Plateau, close to that part of the Indian Tectonic place now known as Pakistan, where the early Lemur bones of *Bugtilemur* were found.

The split between the Seychelles and the Indian plate could also have let to further splits along this link, commencing the creation of the deep-water channels between these islands and between them and Madagascar.

Looking at the seabed map, however, there was never sufficient land above water on this land bridge during this last ice age, for it to be called a continent. The physical land of Lemuria did not exist, except as a pleasant myth.

c) <u>MU</u>

Mu has a checkered history to match that of Lemuria. It was also invented as a name in the 19th century. It came from a mis-translation of Mayan text by Augustus Le Plongeon in the late 1980s[81], where he invented the idea of Mu, and placed it in the Atlantic. The name was taken up by a Colonel Churchward[82], who claimed to have been shown secret tablets. He redefined Mu as covering most of the Pacific Ocean, stretching from Easter Island, to Hawaii and the Marianas.

This was generally described as pure fantasy, until it was discovered that, in the last Ice Age, the lowered sea level had revealed a land mass covering Vietnam, Malaysia, Borneo and Java. This was called Sundaland[83], and fits the description of MU that has been conjured up by the romantics.

Then things took a turn for the fascinating, when dredgers in the strait between Java and Madura turned up a cache of ancient bones, including two skull fragments of skulls of *Homo Erectus*, dating back 140,000 years. This region goes back a very long time.

This was topped by the discovery of a pyramid in west Java, which was originally thought to be dated to about 2,000 BCE. However, when it was investigated using core-drilling and ground-penetrating radar, it was claimed that radiocarbon dating of its lower layer showed it was between 25,000and 27,000 years old[84].

This claim has resulted in open warfare in the archaeological world, because it would predate the pyramids of Giza by nearly 20,000 years. Investigations are on-going.

The Mu enthusiasts have added to this discovery by deciding that Easter Island, Nan Madol[85] in Micronesia, and Yonagui[86] near Japan were all colonies of Mu. UAP enthusiasts have joined in, suggesting it must have been a colony of Ancient Aliens. Best of all are the suggestions that this is the source of Rh-Negative blood on Earth, and theat the Basque race are their direct descendents.

Some of these may turn out to be true but, what is definitely true is that there is a sunken land in the Pacific Ocean, whatever it might have been called then, but proponents have decided to call Mu today.

d) ATLANTIS

i. Introduction

And last, but not least, we come to the great-grand-daddy of them all – Plato's Atlantis. Because it has been around as a concept for far longer than Lemuria or Mu, there has been far more opportunity for speculation and even genuine research. Did Plato invent the whole idea as a cautionary tale, or was it based on folklore or historical events?

Plato describes Atlantis as surrounded by large areas of fertile land and having canals. He says it lies beyond the Pillars of Hercules. As studiers of mythology always seem to enjoy a good argument, they started with whether this referred to the Straits of Gibraltar or somewhere else.

ii. Inside the Straits of Gibraltar

Within the Mediterranean, we have:

1. Fantasy It is a complete fantasy dreamed up by Plato as a cautionary tale.

2. Helike It is based on the destruction of the city of Helike[87] on the coast in the Gulf of Corinth in 373 BCE, during Plato's lifetime. It's destruction does not fit in with the timescale which Plato gives, so could only be cited as an inspiration for his work, rather that the real Atlantis.

3. Turkey It is based on small states on the southern coast of Turkey, namely Troy[88], or Lydia[89]. Both these books concentrate on the destruction of their chosen city-state. The whole area of Anatolia, which includes these states, is known for having developed tin mining, leading to Bronze and then Iron. This was preceded by the development of

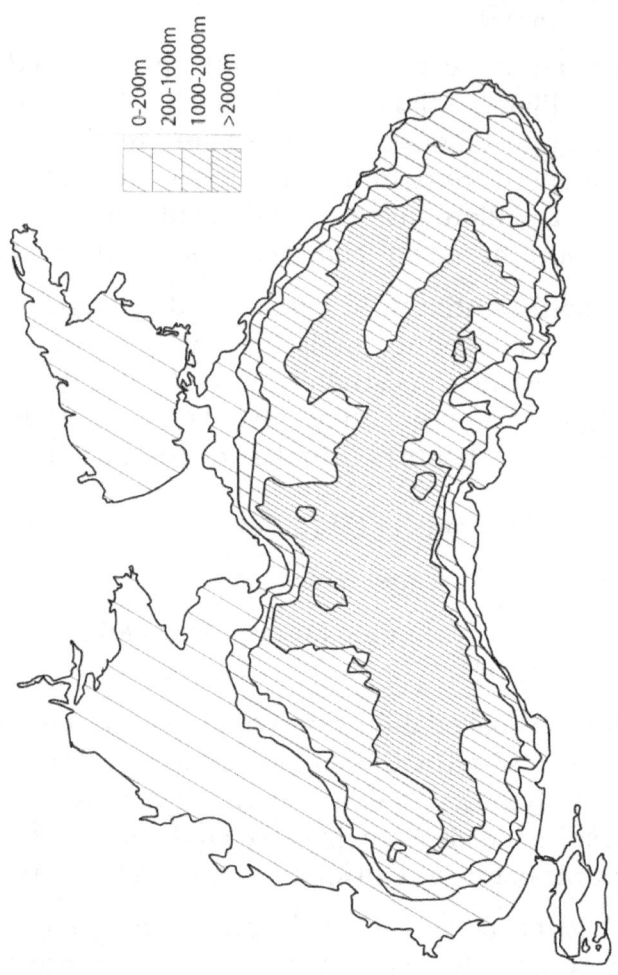

Figure 11 The Black Sea

Agriculture further south east in Mesopotamia. This could have been the cradle of European civilization[90].

4. <u>Black Sea</u> It is suggested that part of the Atlantis story is based on memories of the sea breaking into the Black Sea in about 6,000 BCE, perhaps with embellishments added over the years. (Figure 11) The proponents of this possibility, such as Patrick Chouinatd[91], largely look on the suggestion that Atlantis lies beyond the Straits of Gibraltar as a pointless distraction. They claim that the flooding of the area around the Black sea destroyed a budding civilization, which was advancing faster than surrounding societies.

There is some disagreement over how much lower the Black Sea was then. It is thought to have been between 100m and 140m[92], but there have been suggestions lately that it was more like 30m lower, or that it never happened at all. This doesn't seem sufficient to have had such a major impact on the inhabitants' psyche as history suggests. Also, we would have been more likely to discover the remains of sunken coastal ports under less than 20m of water. The evidence for the greater depth seems overwhelming.

Chouinard claims that the destroyed civilization was a successful farming society which had also mastered stone construction, building ports and public edifices. He suggests they were working copper but had not discovered how to make bronze.

The flood drove this society apart, going in every direction, taking their skills, language and genes with them, possibly giving rise to the concept of Aryan races.

Significantly, many refugees fled south and arrived at Çatalhöyük in central Anatolia, where they stayed.

5. <u>Santorini</u> It is based on the volcanic eruption on the island of Santorini and the destruction of the Bronze Age Minoan Empire in Crete by a tsunami in 1600-1500 BCE[93]. This is not really ancient enough.

6. <u>Cyprus</u> It is based on the civilization in Cyprus[94], which started about 9,500 BCE, but whose cities were also inundated by tsunamis. In fact this book makes a better case for Anatolia in Turkey than for Cyprus. It is more likely that Cyprus became an important trading post, supplying Anatolia with plenteous amounts of copper.

iii.   <u>The Land of Dwarfs</u>

Turkey is surrounded on its three coastlines by alleged sites for Atlantis, even if none of them meet the basic criterion of being outside the Straits of Gibraltar. These are the Black Sea to the north, Troy to the west, Lydia to the north-west and Cyprus to the south.

It has been proposed[95] that there are "Other" bases on Earth which are being used to host the Extra-Terrestrials who are working to raise the levels of morality and technology of Humankind so that we can, one day, qualify to join them in "polite society".

We can look for Atlantis later, but there does not have to be just a single site being used by these "Others". There are at least three locations – Atlantis, Mu in the Far East and Anatolia in Turkey. This last is complete with current archaeological sites, demonstrating its presence, and even the type of "Other" who occupied it (See Figure 12).

52

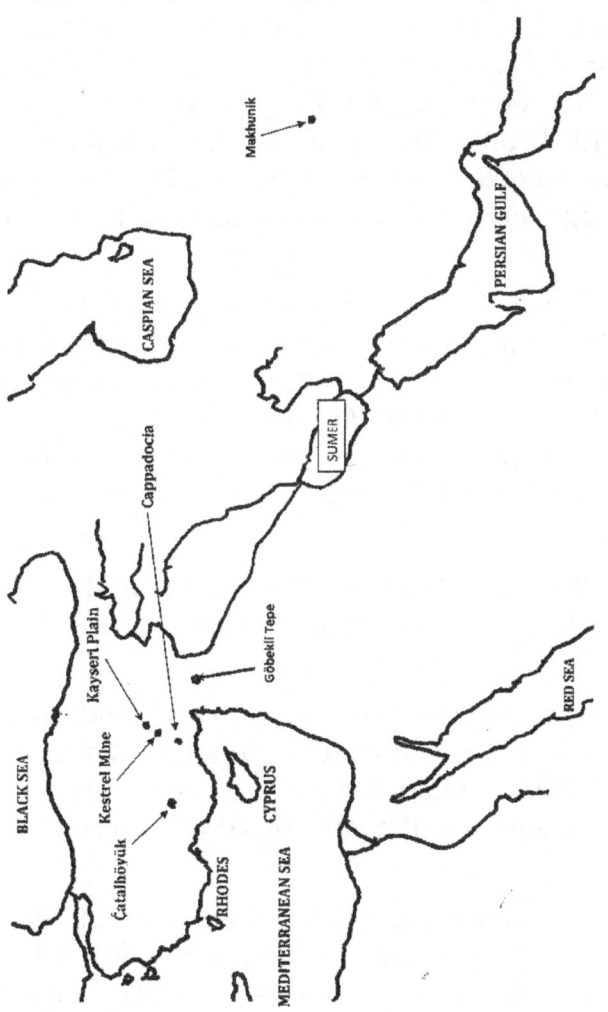

Figure 12 Principal Archaeological Sites In & Around Turkey

A species called the Anunnaki[96] started teaching a group of Humans in the area we know as Sumer. We know this from the clay tablets found there. Sometime before 10,000 BCE a group of Dwarfs came to Anatolia, and starting digging and building homes and mining for tin and copper. They first made Arsenitic Bronze then, once they had found tin[97], proceeded to make Selenitic Bronze as we know it today. This knowledge they passed on to Humans, initiating our Bronze Age.

We know that Dwarfs were involved because of the size of edifices in their underground city at Cappadocia[98], of their Kestrel[99] tin mine near Niğde, and of their tiny homes in Çatalhöyük[100] and Makhunik[101]. The first of these townships, Çatalhöyük, was occupied by Humans following the flooding of the Black Sea area and the Dwarfs moved to Cappadocia.

It is also possible that they were the creators of the structures, believed to be temples, at Göbekli Tepe[102] and nearby.

It has been suggested that the "Others" used children for tunneling and building, but people 1.5m tall[103] have only disappeared from Makhunik in the last hundred years, so there were certainly small people there who weren't children.

It is unlikely that there is a significant presence of Dwarfs in this area today. As I reported in an earlier book[104], they probably believe that their work is done and they have mostly retired to their underground colonies which include those in northern Sweden and Genoa in northern Italy.

Interestingly, though, there is reference[105] to a mysterious group in Turkey called the Dactyls, a class of sorcerers and scientists. They were painted as healers, metalworkers and musicians.

In Turkey's neighbor, Greece, there are reputed to be areas where Little People called Dactyls used to live[106]

Phrygian Dactyls living around Mt Ida in Crete, Kabeiroi Dactyls in Lemnos, Samothrace and Thebes, and Rhodian Dactyls, living presumably in Rhodes.

Is the tale of Dactyls the sole folk memory of Dwarfs in Anatolia?

iv.   Outside the Straits

Then we have the proposed locations in the Atlantic beyond the Straits of Gibraltar, which the proponents claim are the true Pillars of Hercules:

1.   The Azores. These are a group of 9 islands at 30°W x 38°N, on a submerged plateau 300km x400km, 1km deep, in the middle of the Atlantic Ocean. It sits on the junction of three tectonic plates – the Eurasian, African and American, and is still volcanically active. (See Figure 13). It is claimed that, when Cabral first landed there in 1431 CE, there were no Humans or mammals present[107]. However, Otto Muck[108] justifies his belief on three counts:

He believed the Gulf Stream would be reflected rather than simply diverted by an island the size of Spain, so permitting the glaciations of northwest Europe,

He did not try to account for the gap in Gondwana necessary for such a land mass to have been present.

He did not believe that deep water soundings there could be credited to the mid-Atlantic ridge.

In claiming the Gulf Stream would prevent glaciations of north-west Europe, did he allow for the fact that the whole of Earth would have cooled, including the Gulf of Mexico? This would certainly have reduced the power of the current, and the gyre could have collapsed completely.

In general, his proposed Atlantis would have been too big to hide, and too small to have the effects he describes.

2. The Canary Islands[109]. These are sea mounts with very deep waters around them. It is difficult to see how they could possibly have been Atlantis, as there is no evidence of catastrophic destruction. There is probably an "Other" base there today[110], but it is hidden underwater.

3. The Sargasso Sea. A convenient hypothesis, as any physical evidence would by now have sunk below it. There is no evidence to support this hypothesis.

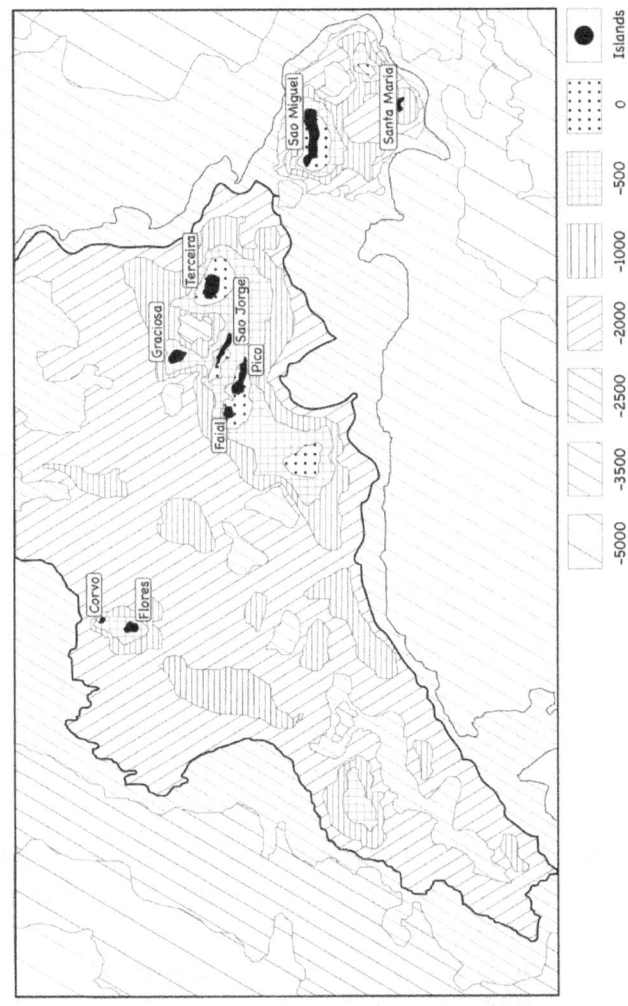

Figure 13 The Azores underwater plateau.

4. The Bahamas. The Bimini Road, a formation of limestone rocks underwater off the island of Bimini, is claimed to be man-made. Naturally this is disputed. An undated anonymous book[111] seems to suggest that is no other evidence available which supports the argument that it is not natural. However, there is now some evidence of man-made structures beneath the sea[112], but within Cuban territorial waters. At present, further investigation is not possible.

5. Cuba. In his book, Andrew Collins[113] makes a strong case for an Atlantis based on Cuba (See Figure 14) and The Bahamas. He spends about half his book demonstrating that the Phoenicians could well have reached the Americas long before Christopher Columbus, and thus Plato could well have known of the West Indies.

    Collins then argues, as did Muck[114], that the Atlantean civilization was destroyed by a comet when it crashed into the Atlantic Ocean north of Puerto Rico about 10,800 BCE, sending massive tidal waves to flood surrounding islands.

    The initial destruction was compounded by a rise in the sea-level caused by a surge of melt-water at the commencement of the Younger Dryas and the start of a mini-Ice Age. Most of The Bahamas Greater and Lesser Banks were permanently submerged. The comet also caused the Carolina Bays as it passed over the US.

    It is possible that the mountains on the north coast of Cuba could have protected cities there, But Humans on the Bahamas Banks would not have stood a chance.

He suggests that the survivors of Atlantis may have headed for the Yucatan peninsula and that the Mayans could be their descendants.

6. South America[115]. The author makes quite a convincing case for there to have been a city near Lake Titicaca, with irrigated fields in what is now the Atacama Desert. As these are in a basin, high rainfall could have lead to this whole area being inundated, and it would have been very slow to drain away. Hence the demise of the city. His statement that there was an Atlantean empire stretching along the north coast of Africa is far from proven, however.

v. An Overview

Basically, people have attempted to place Atlantis just about anywhere where there is the slightest possibility it could have existed, even if there is no supporting evidence.

Often proponents of a particular option seem more interested in rubbishing other options, rather than properly substantiating their own. In doing so, they seem to forget that there could be more than one site in the region which was in use at that time.

Also, they are making gross assumptions about the occupants, assuming them to be a Stone Age society, even though they may at the same time claim them to be high technology. All the evidence which has been found justifies the Stone Age assumption. Any high-technology is notable by its absence.

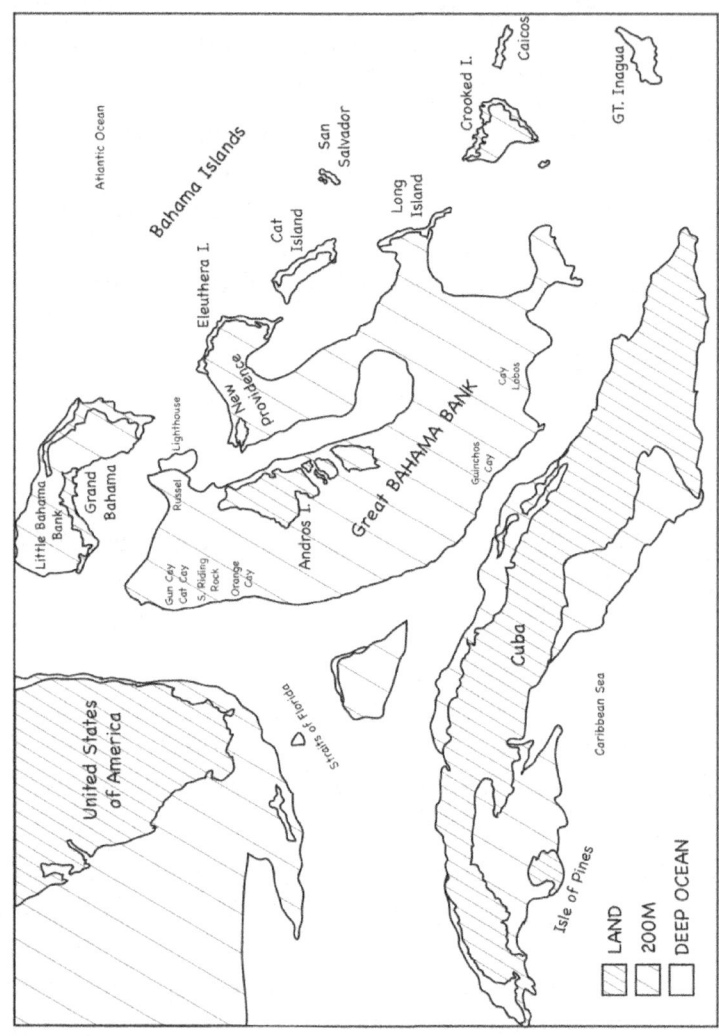

Figure 14 Cuba and the Bahamas

It is my belief, with some justification[116], that the occupants of these sites were "Others" who were working to raise the abilities and morality of Humans. It is highly probably that they were working from a one of a number of sites around the world: Anatolia in Turkey, Mu in South-East Asia and Cuba/ The Bahamas in the Caribbean. There was at least one possible sub-base high in the Andes at Lake Titicaca.

It is possible that the "Other" Caribbean base was on Cuba, and they were working to improve the lot of the inhabitants of both Mesoamerica and the Bahamas Banks. There are certainly indications of Stone Age activity in both places.

Sadly this base was destroyed by a freak comet impact leaving few survivors. For unknown reasons, the "Others" did not re-establish this base to continue their work. For this reason, the inhabitants of Mesoamerica were left to their own devices, stuck in the late Stone Age or early Bronze Age whilst the other regions, Europe and China continued with their tuition. It was not until relatively recently that they started to receive these benefits again.

e) SUMMARY

There appear to have been three "Other" bases present at the end of the last Ice Age – Anatolia in Turkey, Mu in South East Asia and Atlantis in Cuba/The Bahamas. These housed teams working to develop Humans both morally and technologically[117]. At that time they were teaching us the basis of agriculture, and were moving us into the Stone Age.

Despite many claims to the contrary, Lemuria, in the west Indian Ocean, was simply a short-lived land bridge between Madagascar and the Indian tectonic plate over 50 million years ago. This allowed some early Lemurs to reach the area we now know as Pakistan, leaving their fossils for us to find. It cannot meet claims that it was a sunken continent.

Unfortunately, the Atlantis base in Cuba was destroyed when the remains of a comet struck Earth as the Ice Age ended, streaking over what is now the US, and impacting the Atlantic Ocean some distance north of Puerto Rico. This left its marks on what is now Carolina, but caused massive tidal waves in the Ocean, which inundated many parts of the West Indies, apparently destroying the "Other" base of Atlantis. There were few survivors.

The consequence of this was to cause the "Others" to abandon their attempts to civilize the Americas, and to concentrate on China and Europe. Here civilization proceeded to develop through various stages as described in my earlier book[118], whilst Mesoamerica was left in the Stone Age, or perhaps the early Bronze Age, where it remained until it became influenced by the other two developing civilizations.

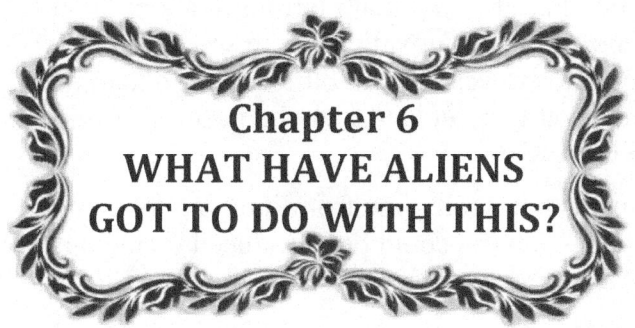

Chapter 6
WHAT HAVE ALIENS GOT TO DO WITH THIS?

a) THE ANCIENT ALIENS STORY

About 12,000 years ago, in Mesopotamia, it all started with a human tribe of hunter-gatherers which was eking out a living on the banks of the Euphrates. Clay tablets[119], found by modern archaeologists, describe how a species of "Others" called the Anunnaki, giant humanoids, possibly assisted by Avians[120], a bird-like species of hominid, arrived there and started introducing them to ways of making their lives easier. The humans learnt agriculture, mathematics, astronomy, writing, schooling and the calendar. They were the first to introduce the 24 hour day and the 60 minute hour. It was also during this time that the wheel was invented.

All these innovations are important, but that of agriculture was crucial. By learning about irrigation, cattle husbandry, the gathering and sowing of seeds and rotation of crops, they could produce more food for the same effort. Alternatively, it provided spare time for other activities.

Pictures of Anunnaki and Avians appear in Egyptian paintings and carvings, and this suggests that they were spreading their tuition widely. They would probably have had dealings with the early Persians, Turks and Greeks as well.

It is claimed that the Anunnaki left Earth about 5000 BCE, by which time all these innovations were spreading around the world. Perhaps it was really a bit longer ago than that, and their departure actually coincided with the flooding of the Black Sea area in 6,000 BCE

Once humans had spare time from scraping together enough food to survive, ideas could be introduced to encourage their curiosity and creative urges.

Copper had been worked for a long time during the Stone Age but, about 3,000 BCE (5,000 years ago, not that long after the Anunnaki left), tin was first discovered in Anatolia in Turkey[121]. Very soon after that, tin was added to Cypriot copper to make bronze, and the Bronze Age began, again in Anatolia. Who thought to do that? It certainly advanced civilization significantly.

Certainly someone like the "Other" race known as Dwarfs, small people famed for their stone and metal working skills, would have had the capability to work with materials like these, and it is possible that they did exactly that.

It seems that "Others" had started to educate Humans in Anatolia, and probably in China and South America.

In Europe, the knowledge of agricultural skills spread to the north-west, initially following the Danube, as well as spreading to Egypt and Greece.

b) WHERE DID THEY LIVE?

The most obvious possibilities are the sunken lands of Atlantis, Mu, Anatolia and the Titicaca basin and Cuba areas. These have the advantage that all signs of their presence have been nearly eradicated. Perhaps there are some indications under the waters, but they would be hard to find after all this time.

One can imagine that the dwelling of the Anunnaki would have seemed awesome, giving rise to some of the magnificent descriptions that have come down to us.

c) WHAT ELSE WERE THE "OTHERS" UP TO?

As I described in Chapter 2, the "Others" were responsible for driving the Reptilians off Earth or into caverns underground. Sadly, since then they have also accidentally given the Reptilians the opportunity to return, when "Others" were forced to respond to our use of nuclear weapons.

In my earlier book[122], I showed that the Fae, the smaller species of "Others", have been on Earth for thousands of years. They don't interfere in the government of Earth, but they do seek to protect the environment, and some species delight in playing tricks on Humans, giving rise to many Ghost and Poltergeist stories.

In Part 3 of my Trilogy[123], I showed that the "Others" have been carefully steering the Human race on the path to higher technology and morality in the hope that we can become acceptable to the other species that abound in the universe.

d) SUMMARY

The Human race has been steered for millennia by "Others" in many different ways. In the beginning, these could have built their own cities, which could have been luxurious and quite miraculous to primitive eyes. These cities have now mainly disappeared beneath the rising seas after the last Ice Age.

These benevolent "Others" also protected Humans from predatory species of "Others", and sought, secretly, to raise our level of civilization.

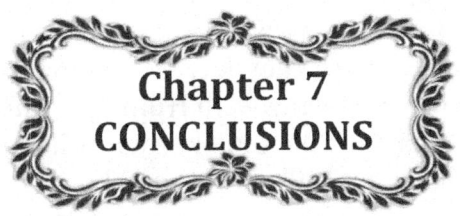

Chapter 7
CONCLUSIONS

In this book, I have been looking at some of the wider aspects of a world where the existence of UAPs and "Others" (whom some refer to as aliens), is taken as proven. There are other anomalous happenings on Earth, which some think require a degree of mystical belief. I have sought explanations which don't.

In my last book, "When the Fae Came", I found that Fairies, Dwarfs, Elves and their like, could be explained without the need for "Fairy Realms" or alternative dimensions. They are simply smaller species of "Others", with many of the same capabilities. They have been here longer than most, and are very widely spread.

In this volume, I have looked at four purportedly fictional genres.

First, I have looked at the supposed presence of reptiles, who claim to own Earth, on the basis that they were here prior to Human emergence. I have shown that they were probably an invading species who, during the last Ice Age, were expelled by "Others" seeking to encourage Human development.

Some of these reptiles, rather than leaving, chose to hide away underground and await the return of their kin. Over at least 10 thousand years, they have developed their own way of life, not really interfering with Humans. These varied species of reptiles have become known collectively as the Saurians. Humans are still threatened by their kin, the Reptilians, however, who are now trying to regain, covertly, their control of Earth.

Secondly, I have been looking at Ghosts, whether apparitions and noises which are either simply echoes of past events, or presences apparently aware of their surroundings. They can be explained in terms of a physical hypothesis called the "Stone Tape Theory", combined with the mischievous actions of number of species of the Fae. However, this hypothesis needs to be further explored.

Thirdly, I looked at the whole range of Sea Monsters reputed to occupy Earth's oceans. I was able to show that some were simply survivors from the extinction event which eliminated the land dinosaurs. It is accepted that snakes, Octopuses, Squid, and Turtles were there before the event and are still here today, though they have had 60 million years to evolve since then, leading to at least one giant - the Colossal Squid.

Whale, which include the largest creatures ever known to have existed on Earth – the Blue Whales at 200 tonnes - are agreed to have evolved since the extinction event, probably from early mammals who returned to the oceans.

More contentious is the claim that Sea Serpents have evolved from snakes, although these could be Oarfish which usually live in the deep oceans. The jury is still out, but who knows what we may find?

Most exciting is the body of evidence which points to a sea dinosaur survivor descended from early plesiosaurs, which has been called Nessie. These are not just occasional residents in Loch Ness, but have been sighted world-wide, particularly in the southern oceans. There is a breeding population of these creatures, who may be using fjords and accessible lakes as nurseries for their young.

Fourthly, it seems likely that, at least in the early days, the "Others" built themselves quite luxurious bases from which to operate, and these probably led to the stories of the magnificent city of Atlantis. The "Others" were not just at a single location. There were at least three sites, one in Turkey, one in Mu in south-east Asia, and one in the Caribbean centred on Cuba and The Bahamas. There may have been a further base up in the Andes in what is now the basin of Lake Titicaca.

Sadly, one of these bases, Cuba/The Bahamas, was wiped out by a comet which caused massive tidal waves. The other two bases have faded away, but could still be operational today.

Overall it would appear that a lot of Human mythology has come from our interactions with "Others" and our miss-interpretation of what we are seeing.

APPENDIX 1

Species of "Others" under 1.5m tall, identified as resident on Earth

Alux	Human figure, only 0.3m tall
Angel	Shining white, 0.5m tall.
Boggart	A disruptive Fae (typically a Brownie) who has fallen out with the house-owner, and plays tricks on them.
Brownie	House Elf 0.75m tall Likes doing house-work at night. Likes to live behind the hearth.
Capelobo	1.2m – 1.5m tall, with a trunk-like. nose. A forest-dweller
Chupacabra	Dog-like vampire 1.0-1.2m. May be only a pet.
Curupira	Feet backwards, goggle eyed, bad acne, red hair 0.60m
Duende	Hairy Goblin Likes to cause trouble in the home
Duergar	Brown Dwarf 0.45m, trouble maker
Dwarf	White Dwarf 1.0m tall. With beard. Known for its mining and blacksmith skills.
Elf	Slender, graceful human-like with pointed ears 1.2-1.65m

Fairy	Tiny winged creature 0.15m. Loves dancing. May be only a pet.
Farfadet	Hairy Pixie but with antennae, so may not be a Pixie at all.. 0.4m
Gnome	Small Dwarf 0.4m. Could be the Knocker heard in mines, warning of danger.
Goblin	Ears stick out 0.6m. Generally hostile to humans.
Hobgoblin	House Goblin 0.45m. Friendly till upset. Enjoys doing house-work at night. Likes to live behind the hearth.
Kappa	Japanese water dweller vampire. Deadly
Karka	Frog-like 0.8m – 1.0m tall, with protruding stomach and buttocks. Walks awkwardly.
Leprechaun	Fae shoemaker 0.75m. Dresses smartly
Lutin	Goblin, 0.45m No permanent home
Menehunes	1.0m Dwarfs. Beardless. On Pacific islands
Michelin Man	Man in space suit 1.2m
Octopus	1.2m With at least 4 arms
Pixie	Small Elf 0.60m loves dancing

Appendix 2

List of Reptile Sightings Worldwide 1900-1999

YEAR	COUNTRY/ STATE & PAGE	DETAILS	UAP? Y/N
1918	Australia 55	10m long lizard seen in mountains	N
1921	Austria 16	Hole-dwelling lizard 1m 4 short legs with 3 toes, forked tongue. Springs to attack.	N
1932	USSR (W) 22	Bipedal dinosaur. Large head, bulging eyes, trapped in ice	N
1936	USSR (W) 29	1.7m creature, scaly, with tail and legs	N
1937	Chile 182	Pink aquarium of fish & reptiles on lorry	N
1938	Estonia 50	Frog-like hominid. 1m brown/green long legs, wide mouth	N
1939	France 45	2m Saurian seen in Pyrenees	N
1943	Belarus 14	Short reptilian humanoid with fish-like scales, webbed fingers, entered lake	N
1949	USSR (C) 47	1/0m, snake head, comb on back, hostile	Y

YEAR	COUNTRY/ STATE & PAGE	DETAILS	UAP? Y/N
1951	Illinois	0.45m frog-like creature, striped skin, 3 toes, slit mouth, wrap-around eyes	Y
1952	Ukraine 208	8m reptile, plates on green/brown skin, claws, snake head, green eyes	N
1953	Australia 84	2.0m scaly body	N
1953	Cuba 62	Cross Iguana/Chameleon/ Alligator, human arms & legs, claw-like nails	N
1953	Denmark 41	Cold, rough scaly skinned humanoid	N
1954	Spain 42	1.5m lizard on 2 legs, dark, with tail	N
1954	France 96	1.2m green, large eyes, tail	N
1954	Australia 89	2m scaly back, 2 flippers, long tail	N
1954	New Zealand 95	2m scaly body. Near Barri	N
1954	Italy 134	1.3m scaly like a fish	Y
1955	Ohio	Frog-like, grey skin, thin line for mouth, large chest, one arm longer than other. Rolls of fat across head.	N
1957	England 111	Brown/green scaly skin, black eyes	Y
1958	California 90	1.8m Scaly body, round head, protruding lips	N
1959	Hawaii 118	Green lady, scaly skin, long claws	N

78

YEAR	COUNTRY/ STATE & PAGE	DETAILS	UAP? Y/N
1960	Argentina 33	Giant lizard man in sewage tunnel	N
1960	Turkey 304	2m reptile	N
1960	New Mexico 147	1.2 reptilian encountered in tunnel	N
1962	New Jersey 168	Tall reptile	Y
1963	N Carolina 68	Lizard dressed in toga	N
1965	Mexico 114	Tall, scaly toad-like mouth. Hands like flippers	N
1965	Mexico 116	Short reptile, long tail	Y
1965	Brazil 75	Man-like greenish scaly	N
1966	Indiana	Dark scaly creature, red eyes, one arm longer tat the other.	Y
1966	Kentucky 89	1.7m brown scaly amphibian, webbed feet & hands, dark eyes, gills	N
1966	Mexico 116	2m scaly reptile. Dark orange, webbed fingers.	Y
1967	Spain 59	0.7m Frog-like being	N
1967	Alberta	2m monster, brown scaly skin, holes for ears and nose, 4 fingers	Y
1967	Florida 42	1m red fish scales. Face, no neck	N
1968	Spain 65	2.2m lizard, red eyes, claws, tail big tongue	Y
1969	Uruguay 286	Tall lizard-like. Green scaly skin	Y
1969	Colorado 61	Thin, green, scaly, Lizard-like, webbed hands V-shaped feet	Y
1969	Brazil 112	Tall man-like green scaly	N
1969	Australia 185	Skin like a frog. Big eyes	N

79

YEAR	COUNTRY/ STATE & PAGE	DETAILS	UAP? Y/N
1970	Italy 167	5m lizard with thick legs	N
1970	Iowa	1.8m scaly creature, large head, broad shoulders	N
1970	Uzbekistan 309	2m scaly slimy skin, snake-like face, big eyes, arms & legs	N
1971	Wyoming	1.1m leathery skin, frog-like orange eyes, small ears and mouth	N
1971	Arizona 24	2m, skin like crocodile, 3 fingers, foot like elephant	Y
1972	Mexico 128	Reptile/Human hybrid from lake	N
1972	Puerto Rico 202	Tall & Short reptiles green/brown	N
1972	Japan 149	1m starfish with frog-like skin	Y
1972	Puerto Rico 203	1.2m thin, scaly, claws tail	N
1972	France 166	Green scaly skin - Abduction	Y
1972	Ohio	1m leathery skin, frog-like face.	N
1973	Argentina 89	1.7m thin reptile, protruding dorsal fin, short tail, clawed feet	N
1973	New Jersey 165	Giant man-like alligator	N
1973	Puerto Rico 205	Scaly Parrot-like nose	N
1974	England 283	Massive toad-like entity	Y
1974	Argentina 93	2m with scaly skin	N
1975	Italy 176	3m lizard, thick as a dog, wolf howl	N
1976	Washington 256	Huge reptilian, wide slit mouth, bulging eyes, went into lake	N
1976	Dominica 77	1.5m Green scaly arms	N
1976	Michigan	Large reptilian with huge face, wide slit mouth, bulbous eyes, flipper hands, jumped into lake	N
1976	Spain 116	2.1m beings with crest along back and tail	N
1976	Philadelphia 110	Scaly skin, cat-like face	Y
1976	Spain 127	0.4m reptilian "penguin", strong jaws and claws. Captured	N
1976	Puerto Rico 215	Underwater, fish-like scales, face	N

80

1976	France 203	1.2m green, webbed hands & feet red eyes	N
1976	Spain 125	1.6m scaly humanoid, prominent nose, thick upper lip	Y
1976	Michigan	1.5m green, pincers, no eyes, ears, with skin bumpy like an alligator.	Y
1977	England 351	1.5m green-scaled reptile	N
1977	Puerto Rico 216	1.2m Grey/green scaly	N
1977	Brazil 156	Lizard-skinned vampires	N
1977	Peru 272	1m green scaly 3 fingers	N
1977	England 342	1.5m green scaled, pointed ears, red slanted eyes, filthy teeth, 4 fingers, tail.	N
1977	Puerto Rico 217	1.1m scaly, with tail, duck feet	Y
1978	Argentina 107	1.1m green. Claws on hand when glove removed Abduction	Y
1978	Philadelphia 42	1.2m Scaly	Y
1978	Ontario	Upright alligator	N
1978	Japan 154	3m tall reptile with thick scaly skin and yellow eyes. Came out of sea.	N
1978	Mexico 146	1.5m Green scaly skin, claws	Y
1978	Italy 212	Tall green reptilian	Y
1979	USSR (E) 121	Monstrous Reptile - abduction	Y
1979	Finland 86	Little, toad-like skin with green/brown spots, bulging eyes, pointed ears, hawk –like nose	Y
1979	Florida 54	Humanoid, scaly, large oblique eyes	Y

YEAR	COUNTRY/ STATE & PAGE	DETAILS	UAP? Y/N
1980	Brazil 127	Men changed shape to look like reptiles with scaly heart-shaped heads	Y
1980	Spain 163	Frog-like green, scaly round head	N
1981	Brazil 182	Cross between frog and reptile	Y
1982	Australia 302	1.6m lizard, large eyes, scaly hands with claws - abduction	N
1983	Missouri 132	Tall green lizard with vertical pupils, webbed hands and feet	Y
1983	USSR (E) 127	1.45m green frog/reptile, 4 webbed fingers. Large head, huge mouth no neck	N
1984	Chile 227	Tall lizard man, long tail, big head	N
1984	Germany 109	Creature with fish-like scales	Y
1984	California 178	Frog-like being	Y
1984	Puerto Rico 244	Upright lizard, no tail 1.2m yellow	N
1984	Mexico 154	Giant lizard emerged from rock	N
1984	Ukraine 231	1.6m scaly lizard muscular arms and legs with claws, green slanted eyes, vertical pupils	N
1992	Uzbekistan 316	Snake-like entity	N
1984	England 471	Small reptilian creature	Y
1985	Italy 223	2m reptilian, webbed feet, red eyes	N
1985	USSR (S) 153	Small reptilian, gray, 4-fingered webbed hands	Y
1985	Kazakstan 155	Reptilian, small wings, dark eyes, pointy nose	Y
1986	USSR (C) 165	Reptilian – abduction & rape	Y
1986	Italy 225	Tall green reptilian red eyes	N
1987	Puerto Rico 250	1.2m scaly humanoid	Y
1987	Italy 227	Huge scaly reptile. Big head, red eyes	N
1987	Puerto Rico 250	Very large iguana, bat-like wings, tail	N
1987	Connecticut 20	Giant Lizard man	Y
1987	N Carolina 77	Man-like, scaly, tall small head	N

82

1988	USSR (S) 175	1.0m reptilian, blue/green, large head, 3 webbed fingers. Black eyes with eyelids	Y
1988	Italy 228	2-3m scaly humanoid, snake-like eyes	N
1988	England 498	1.7m lizard-like with long head. Abduction and rape	Y
1988	Missouri 134	2.2m reptile	Y
1989	Texas 240	Large lizard-like beings, abduction	Y
1989	Ontario	Reptilians killed with poison gas	Y

YEAR	COUNTRY/ STATE & PAGE	DETAILS	UAP? Y/N
1990	Puerto Rico 263	Avian reptile, pointed beak, horn on head	N
1990	Scotland 150	Reptilian entity	Y
1990	Argentina 154	Green, wide mouth, big black eyes, rough skin	N
1991	Poland 217	Tall with scaly horned body, a bit like a dragon	N
1991	Missouri 135	1.8m reptile, green eyes, slanted pupils	Y
1991	USSR (W) 327	6-7m Snake, 0.35m thick, green in lake	Y
1992	Latvia 155	Bodies of frog-like hominids, 1.3m long thin arms and legs with webbed fingers yellow/green/brown	Y
1992	Vietnam 330	Human body, skin like lizard	N
1992	Lithuania 182	Large frog-like creature antennae	N
1992	Italy 231	1m. White skin with scales	N
1993	Brazil 217	Tall, green scaly skin, huge head, large eyes light green	N
1995	Ukraine 301	2m powerfully built reptile, long arms with claws	N
1995	California 189	Tall Reptilians	Y
1995	Missouri 135	2.1m brown scaly skin, flat nose, no ears, ridges on head. 1.5m tail	N
1995	Malaysia 209	0.8m green scaly-skin frog-like eyes	N
1995	Florida 65	1.5m lizard-like, pointy head 1m tail	Y
1995	Ukraine 301	2m blue reptile with clawed hands	Y
1996	Argentina 169	Half man, half lizard in river.	N
1996	Malaysia 209	0.8m scaly body, big head	N
1996	Mexico 177	1.2m lizard scaly skin, claws, spines down back	N
1996	Quebec	Reptilian, long face, with scaly, leathery skin, black eyes, clawed hands, gaunt arms and legs	Y

1996	Palestine 233	2m green scaly skin, frog-like eyes	Y
1996	Romania 273	Gray, big eyes, scaly like fish	Y
1996	Quebec 142	Scaly skin huge arm, legs & eyes, clawed hands	Y
1997	Florida 68	Tall, dark, reptilian	Y
1997	Brazil 231	1.7m pointed ears, scales, green crest over head and down back	N
1997	New Mexico 178	1.6 light blue reptile, claw-fingered toes, yellow eyes, cal-like irises	Y
1998	Washington 286	1m, scaly skin, dark round eyes	N
1999	Oregon 224	Huge reptilian underground	N
1999	USSR (W) 409	Lizard-like. Raped witness	Y
1999	Finland 97	Lizard-like with huge eyes.	Y
1999	Argentina 174	Huge winged half human, half reptile with 2 horns	N
1999	British Columbia	2m tall hominoid with scaly skin. Heavy build	N
1999	Mexico 182	Green, scaly. Crest on back	N

85

APPENDIX 3

Table   List of "Others" over 1.5m Tall sighted on Earth

NAME	HEIGHT (m)	DETAILS
Agathan	2.0-2.6	Earth Human, also known as Telosian
Alcyone Pleiadians	1.6-1.9	Nordic
Altarians	2.0-3.0	Human-like Blue/green/tan skin
Andromedans	1.7-2.1	Blue skinned
Antarians	2.2-2.8	Human-like
Apunians	2.1-2.8	Nordic
Arcturians	3.0-4.0	Blue skin
Arians	1.7-2.1	Nordic
Cassiopians	1.9-2.5	Human-like. Webbed fingers
Certans	1.9-2.5	Human-like
Clarions	1.5-1.7	Petite human-like, slanted almond eyes
Cyclops	1.9-2.0	One eye. Cyclops have also been reported at less that 1.5m tall, so there may be more than one species.
Cygnus Alpha	2.2-3.0	Tall human-like
Eridaneans	2.0-2.2	Nordic. Light blue skin
Itipurians	1.8-2.0	Thin humans
Lyrians	2.0-3.0	Feline

Klermers	2.5-3.2	Tall human-like
Koldashans	1.7-2.0	Human-like
Lady of Light	1.6-1.8	Fair human. Skin glows
Lyrans	2.0-3.0	Human-like
Mantis	2.3-3.2	Insect. Triangular head
Melchizedeks	1.7-2.0	Human-like
Mothman	2.0-2.6	Dark fur, big wings.
Orions	3.3-3.8	Human-like
Pleidians	1.8-2.1	Nordic Fair graceful, slender
Procyonians	2.0-2.2	Feline
	1.8-2.8	Human-like
Proxima Centaurians	1.6-1.7	Human-like
Renegade Pleidians	2.6-4.2	Massive human-like
Reptilians	2.0-3.0	Scaly skin (Extra-terrestrial)
Sagittarians	2.5-4.0	Human-like
Saurians	1.0-2.2	Scaly Skin (Intra-terrestrial)
Sirians	2.0-2.7	Blue skin gills webbed hands
Soulzars	3.0-5.0	Hairless hybrids. Bald
Sylphons	1.2-1.8	Gossamer wings
Tall Whites	2.1-3.1	Slender body
Thurbans	2.1-2.7	Reptilian

Titan Sirians	2.5-4.0	Blue skin, elongated head pointed chin
Unmites	2.5-2.8	Nordics
Vegans	1.8-2.2	Human-like
Venusians	2.0-2.8	Elegant, glowing skin (Valiant Thor)
	1.8-2.1	Human-like

Appendix 4

Sightings of Intelligent Spirits and Poltergeists in the British Isles.

PLACE	DESCRIPTION
Leeds	Kirkstall Abbey. Intelligent Spirit Mary Thwaite, who regrets informing on her husband
Hull	Osborne St. Black Lady, Intelligent Spirit known to scratch, bite and taunt onlookers
Hull	Ye Olde Black Boy. Intelligent Spirit Lady in burlesque costume molests customers, and a pair of hands throws bottles
Pontefract	30 East Drive. Poltergeist
Burton Agnes	Attacking guests and servants Poltergeist
Worksop	Lion Hotel Poltergeist
Chester	Coach & Horses. Intelligent Spirit talked to barman & Drinkers
Shropshire	Three Tuns Inn. Intelligent Spirit talked to barman & Drinkers
Leek	Roebuck Inn Poltergeist
Nuneaton	Griffin Inn Poltergeist
Oundle	Talbot Hotel Poltergeist
Suffolk	Bentley Poltergeist
Suffolk	Potsford Wood Poltergeist

PLACE	DESCRIPTION
Bicester	Holt Hotel. Intelligent Spirit opens doors for beautiful women
Besseleigh	Intelligent Spirit Woman in the woods waves to passers-by
Borley	Rectory Major Poltergeist activity since 18th century
Beaconstree	Underground Station Poltergeist
Colchester	St Andrew's Church Poltergeist
Hertfordshire	Baldock Poltergeist
London	George Inn. Poltergeist
London	White Horse. Intelligent Spirit Woman speaks to people
London	Highgate Cemetery. Dark Intelligent Spirit knocked man over
Colnbrook	Ostrich Inn Intelligent Spirit touches people
Windsor	Black Horse. Intelligent Spirit pinches women
Farnham Common	Crown Inn. Poltergeist, partly exorcised.
Maidenhead	Maiden's Head, formerly The Hobgoblin. Poltergeist
Beenham	Picklepythe Lane Poltergeist
Bracknell	Caesars Camp Poltergeist

PLACE	DESCRIPTION
Avebury	Red Lion Poltergeist
Highclere	Castle Poltergeist
West Kennet	Long Barrow Poltergeist
Somerset	Knowle Poltergeist
Somerset	Nunney Poltergeist
Yeovil	Railway Centre Poltergeist
Leinster	Killmainham Goal Poltergeist
Fethard	McCarthy's Bar Poltergeist
Killarney	Lake Hotel 13yr old girl, Intelligent Spirit aware of other people
County Down	Grace Neill's Bar Poltergeist
Belfast	Golden Tread Gallery Poltergeist
Kirkaldy	Fenars Arms Poltergeist
Glasgow	Greenbank Garden Intelligent spirit which seduces and scratches people
Edinburgh	Greyfriars Kirkyard Poltergeist
Edinburgh	Learmont Hotel Poltergeist
Roxburgh	Jedburgh Castle Poltergeist
Dumfries	Globe Inn. Intelligent Spirit, can tug people's clothing
Cardiff	Mower Service owner. Followed by Poltergeist

PLACE	DESCRIPTION
Tondu	Skeletal Intelligent Spirit picks fight
Abergavenny	Skirrid Mountain Inn Poltergeist
Raglan	Castle Intelligent spirit of librarian
Tredegar	Wife is beaten at night by Poltergeist
Gwent	Tredegar House Intelligent Spirit talks to visitors
Vale of Clywd	Blue Lion Poltergeist
Carmarthen	Llnageler farm Poltergeist
Kidwelly	Mason's Arms Poltergeist

APPENDIX 5

Number of "Other" sightings, in the different areas of the British Isles, comparing beings under and over 1.5m

Mythical Monsters, Sunken Continents & Aliens

REGION	UNDER 1.5m	OVER 1.5m	DATES	% of Total
SOUTH WEST	19	15	Up to 1945	45%
	11	18	1946-1975	60%
	16	11	1976-1999	40%
SOUTH EAST	4	3	Up to 1945	43%
	6	21	1946-1975	80%
	10	15	1976-1999	60%
LONDON	3	6	Up to 1945	66%
	1	3	1946-1975	75%
	12	7	1976-1999	36%
EAST of ENGLAND	8	8	Up to 1945	50%
	5	5	1946-1975	50%
	4	12	1976-1999	70%
EAST MIDLANDS	8	3	Up to 1945	28%
	1	5	1946-1975	83%
	5	2	1976-1999	30%
WEST MIDLANDS	1	1	Up to 1945	50%
	8	11	1946-1975	58%
	9	6	1976-1999	40%
IRELAND	23	7	Up to 1945	23%
	7	3	1946-1975	30%
	15	6	1976-1999	28%
NORTH WALES	5	4	Up to 1945	44%
	4	3	1946-1975	43%
	8	10	1976-1999	55%
SOUTH WALES	4	4	Up to 1945	50%
	1	2	1946-1975	67%
	3	8	1976-1999	73%
YORKSHIRE	4	3	Up to 1945	43%
	2	10	1946-1975	83%
	16	14	1976-1999	71%
NORTH EAST	0	3	Up to 1945	100%
	4	2	1946-1975	33%
	10	1	1976-1999	10%
NORTH WEST	7	6	Up to 1945	46%
	10	12	1946-1975	55%
	12	23	1976-1999	92%
SCOTLAND	12	4	Up to 1945	25%
	6	2	1946-1975	25%
	15	8	1976-1999	35%
TOTALS	98	67	Up to 1945	40%
	66	101	1946-1975	60%
	135	91	1976-1999	43%
	299	265		

APPENDIX 6

Reports of Sea Serpent sightings around the world in the twentieth century.

Year	Place	Details
1903	Scotland	35m Horse head
1903	South Africa	35m Sea Serpent
1905	Brazil	Sea Serpent
1906	England	6m visible
1907	Indian Ocean	Giant eel
1908	Vietnam	Sea Serpent
1910	Scotland	Horse head. Chasing whales
1910	Mid Atlantic	35m Sea Serpent
1911	South Africa	35m Sea Serpent
1912	England	8m long 0.3m thick
1912	Greece	10m Sea Serpent
1916	Surinam	25m Sea Serpent
1923	Cape Horn	Sea Serpent
1924	South Africa	15m Sea Serpent stranded
1925	New Caledonia	12m Sea Serpent Horse head
1930	Australia	25m Sea Serpent
1931	England	Sea Serpent
1933	Scotland	11m Horse head
1934	Bahamas	18m Sea Serpent
1934	Italy	Horse head
1939	Scotland	Horse head
1939	France	4m Sea Serpent
1945	Sweden	Giant Sea Serpent 3 humps
1958	Brazil	Sea Serpent
1961	Ireland	4m Horse eel
1964	Australia	23m monster with big head
1969	Mexico	10m Sea Serpent washed ashore

APPENDIX 7

Recorded sightings of Dinosaur-type Sea Monsters in the 20th Century.

Year	Place	Details
1900	Scotland	Neck up, 6m under water
1902	Peru	In river, 3.5m neck, flippers, 11m long
1905	Cape Horn	1.5m neck
1907	England	Long neck
1907	India	12m body, 3m neck 0.5m in diameter
1912	Gabon	4m neck
1914	Ireland	Long neck, flat head
1917	Iceland	6m neck
1919	Scotland	Head 1.6m, body 3.5m, 4 paddles, dog head
1919	Scotland	Entering water
1920	Brazil	10m swan-like neck
1922	South Africa	12m body, 1.5m neck, 3m tail, fighting whales
1923	Scotland	Crossing road, 6m body & tail, dog head
1925	Australia	Body 8m, fins 3m, tail
1932	Scotland	12m, fighting something underwater
1933	Scotland	Big and small creatures together
1933	Scotland	On beach. 8.0m, short legs, used like a seal. Back like an elephant.
1934	England	Long neck and flippers
1934	Scotland	Long neck, tail 1.6m, body 6m, head flat on top, walked on flippers
1937	Scotland	Two heads sighted
1943	Gulf of Mexico 86	Long necked creature
1946	Scotland	Body size of elephant, long neck
1947	South Africa	Barrel body, head up, big fins
1949	England	Two 6m monsters in the sand dunes
1953	Scotland	3m neck, 10m body
1955	Venezuela	Three 1m long plesiosaurs, sunbathing on the side of a river
1955	Ireland	Long neck, ears/horns, on land
1961	Scotland	Animal with flippers
1965	Sweden	Big, stumpy legs/fins, round head
1969	Mexico 122	Black creature, wading ashore, 2 horns
1970	Scotland	Two creatures seen together

1972	Scotland	Sonar saw two creatures 8m, 2.5m tail, flippers
1974	Scotland	16-20m with a tail
1975	Wales	Monster with flippers
1976	England	7m lizard-like sea monster, long neck, several tonnes

APPENDIX 8

Sightings of Turtles

Year	Place	Details
1883	New Zealand	Turtle 20m x 13m
1905	Brazil	Head like Turtle 2m neck
1906	Sweden	20m, head like Turtle
1921	India	Gigantic Tortoise, 20m
1934	Australia	Head like Turtle
1959	Scotland	Tortoise head

INDEX

REFERENCES

[1] The UFO, ET, Alien Trilogy by Martin Thomas Self-Published 2025
[2] Earth's Alien Syllabus by Martin Thomas. Self-published 2025
[3] Sky People by Ardy Sixkiller Clarke Published by New Page Books 2015
[4] UFOs, Humanoids & Strange Phenomena of Africa, Asia and the Middle East p 199, by George Mitrovic
[5] We Own 29% - ET Has the Rest by Martin Thomas. Self-published 2025
[6] Letters From Mesopotamia
[7] UFO – Friend or Foe Martin Thomas Pub 2025
[8] UFOs: Few answers at rare US Congressional hearing https://www.bbc.co.uk/news/world-us-canada-61474201
[9] Files released on 1974 'Welsh Roswell' https://www.bbc.co.uk/news/uk-wales-10863645
[10] The Fairy Census 1 – Britain & Ireland by S R Young PWCA Books 2023
[11] The UFO, ET, Alien Trilogy by Martin Thomas Self-Published 2025
[12] The Reptile Ascendancy by Fufu Fang. Self-Published. Undated
[13] Reptilians by Ana Lorens. Self-Published 2025
[14] Lizard People by Steve Preston. Self-Published 2018
[15] Amazing UFO Encounters of West Coast North America George Mitrovic, Self Published, Undated
[16] Amazing UFO Encounters of The South GeorgeMitrovic, Self Published, Undated
[17] Amazing UFO Encounters of East Coast North America Pt1 George Mitrovic, Self Published, Undated
[18] Amazing UFO Encounters of East Coast North America Pt2 George Mitrovic, Self Published, Undated
[19] Amazing UFO Encounters of Central North America Pt1 George Mitrovic, Self Published, Undated
[20] Amazing UFO Encounters of Central North America Pt2 George Mitrovic, Self Published, Undated
[21] Amazing UFO Encounters of Canada George Mitrovic, Self Published, Undated
[22] UFOs, Humanoids and Strange Phenomena of Central America, the Caribbean, and Mexico as well as the Atlantic Ocean, George Mitrovic, Self- Published, Undated

[23] UFOs, Humanoids and Strange Phenomena of Bolivia, Brazil, Columbia, Ecuador, Guyana, Suriname and Venezuela, George Mitrovic, Self-Published, Undated

[24] UFOs, Humanoids and Strange Phenomena of Argentina, Chile, Paraguay, Peru and Uruguay, George Mitrovic, Self Published, Undated

[25] UFOs, Humanoids and Strange Phenomena of Russia, George Mitrovic, Self Published, Undated

[26] Amazing Encounters with Monsters and other Mysteries of Australia, New Zealand , the Pacific and Antarctica, George Mitrovic, Self Published, Undated

[27] UFOs, Humanoids and Strange Phenomena of the Ukraine, Belarus, Moldavia and Scandanavia, George Mitrovic, Self Published, Undated

[28] UFOs, Humanoids and Strange Phenomena of France, George Mitrovic, Self Published, Undated

[29] UFOs, Humanoids and Strange Phenomena of Andorra, Gibraltar, Spain and Portugal, George Mitrovic, Self Published, Undated

[30] UFOs, Humanoids and Strange Phenomena of Italy and eastern Europe, George Mitrovic, Self Published, Undated

[31] UFOs, Humanoids and Strange Phenomena of Austria, Belgium, Estonia, Germany, Holland, Kalingrad, Latvia, Lithuania, Luxembourg, Poland & Switzerland, George Mitrovic, Self Published, Undated

[32] UFOs, Humanoids and Strange Phenomena of Ireland, Scotland and Wales, George Mitrovic, Self Published, Undated

[33] UFOs, Humanoids and Strange Phenomena of England, George Mitrovic, Self Published, Undated

[34] UFOs, Humanoids and Strange Phenomena of Africa, Asia and the Middle East, George Mitrovic, Self Published, Undated

[35] Encyclopedia of Alien Races, Maestà Pastore, Amazon, Undated

[36] The Underground Atlas, Middleton & Walton. Pub Robert Hale 1986

[37] Encyclopedia of Alien Races, Maestà Pastore, Amazon, Undated

[38] When the Fae Came by Martin Thomas, self-published 2015.

[39] Encyclopedia of Alien Races, Maestà Pastore, Amazon, Undated

[40] Daily Mail. "President Eisenhower had three secret meetings with aliens" former Pentagon consultant claims. 16th Feb2012

[41] Lizard People by Steve Preston. Self-Published 2018

[42] A History of Ghosts, Spirits and the Supernatural, Pub DK 2024
[43] The Encyclopaedia of Fairies in World Folklore & Mythology by Theresa Bane pub McFarlane & Co 2013
[44] British Ghosts and Where to Find Them, by Nick Tyler, published by Pepperfish 2024.
[45] Ghost Tales of the United Kingdom by Sean McLachlan. Pub Charles Rivers Editors undated
[46] The Roadmap of British Ghosts by Ruth Wylde Self Published 2019
[47] The UFO, ET, Alien Trilogy by Martin Thomas Self-Published 2025
[48] When the Fae Came by Martin Thomas, self-published 2015.
[49] The UFO, ET, Alien Trilogy by Martin Thomas Self-Published 2025
[50] A Natural History of Sea Serpents by Adrian Shine. Pub Whittles 2024
[51] The Princeton Field Guide to Mezozoic Reptiles P64-65 by Gregory Paul, pub Princeton University 2022.
[52] New Scientist, 4th November 2006 presentation to the Society of Vertebrate Palaeontology in Ottawa in the previous month
[53] The Princeton Field Guide to Mezozoic Reptiles P64-65 by Gregory Paul, pub Princeton University 2022.
[54] Nature Magazine 12 August 2015
[55] Consciousness of octopuses—on their own terms. Jenifer Mather *Animal Sentience 2025*
[56] Octopuses p38, by Erica Edwards, Pub 2025
[57] Ibid
[58] Globsters by Michael Newton P38 CFZ Press 2012
[59] The Search for the Giant Squid, by Richard Ellis. Pub Penguin Group 1999
[60] The Deep Sea Phantom by Serena Quillan, Deep Vision Media 2025
[61] Globsters by Michael Newton, Pub CFZ Press 2012
[62] Cryptid Sea Monsters, by Kelly Milner Halls. Sasquatch Books. 2025
[63] The Princeton Field Guide to Mezozoic Reptiles P64-65 by Gregory Paul, pub Princeton University 2022.
[64] The Naked Ape by Desmond Morris Pub Jonathan Cape London 1967
[65] Man and Dolphin: Adventures of a New Scientific Frontier Lilly, Dr John C. Lilly 1961
[66] Britannica

[67] Lost Civilisations P302 by Jim Willis Visible Ink Press 2019
[68] Mammals of Madagascar by Phillip Sclater, Quarterly Journal of Science 1864
[69] Plate Tectonics – a Very Short Introduction by Peter Molnar, Oxford University Press 2015
[70] Legends of Lemuria, the Vanished Civilization, by Barbara White, Self-Published, undated.
[71] Ibid
[72] In Search of Mu by James Churchward, Pub Rudge 1926
[73] Legends of Lemuria, the Vanished Civilization, by Barbara White, Self-Published, undated.
[74] Lemuria Theory of a Prehistoric Continent by Landon Delaney. Pub Xspurts 2005
[75] The Sloth Lemur's Song P 147 by Alison Richard pub William Collins 2022
[76] The Rotating Earth and Plate Tectonics by Robert Maurer. Self-Published 2022.
[77] The Sloth Lemur's Song P 79 by Alison Richard pub William Collins 2022
[78] Sea Level in the Past 200,000 Years. University of New Orleans https://courses.ems.psu.edu/earth107/node/1496
[79] Plate Tectonics – a Very Short Introduction by Peter Molnar, Oxford University Press 2015
[80] The Rotating Earth and Plate Tectonics by Robert Maurer. Self-Published 2022.
[81] The Empire of the Sun: Unveiling the Lost Continent of Mu by Dr D Basak, Self Published No date.
[82] Ibid
[83] Ibid
[84] Ibid
[85] UNESCO world Heritage Convention. The List
[86] Britannia
[87] Greek Reporter https://greekreporter.com/2025/07/15/ancient-helike-earthquakes-Greek-city-rise-fall-return/
[88] The Flood from Heaven by Eberhard Zangger. Pan Books 1993
[89] The Sunken Kingdom by Peter James Pub Jonathan Cape 1995
[90] The UFO, ET, Alien Trilogy by Martin Thomas Self-Published 2025

[91] Atlantis Rising by Patrick Chouinard, National Renaissance Books 2025

[92] Assessment of Black Sea Water Level Fluctuations Since the Last Glacial Maximum. 7 Authors. A Special Paper of the Geological Society of America, January 2011.

[93] Greek Reporter https://greekreporter.com/2025/03/15/minoan-civilization-end-giant-tsunamis/

[94] Atlantis, the Cypriot Empire, Jordi Harth. Published 2024

[95] The UFO, E.T, Alien Trilogy (Part 3) by Martin Thomas. Self published 2026

[96] Scientific & Esoteric Encyclopaedia of UFOs, Aliens & Extraterrestrial Gods V4, p154 by Maximillian de Lafayette Pub New York 2014

[97] Bronze Age Source of Tin Discovered, The University of Chicago Chronicle, 1994

[98] TheCultureTrip.com The Story Behind the Underground Cities of Turkey

[99] Bronze Age Source of Tin Discovered, The University of Chicago Chronicle, 1994

[100] https://whc.unesco.org/en/list/1405/

[101] https://parsdiplomatic.com/iran-sightseeing/makhunik/

[102] Empires of Anatolia Anon, Self Published, Undated

[103] https://www.bbc.co.uk/travel/article/20180109-irans-ancient-village-of-little-people

[104] The UFO, ET, Alien Trilogy Part 2 by Martin Thomas Self-Published 2025

[105] Empires of Anatolia Anon, Self Published, Undated

[106] Giants and the Little People V2 P198 by Peter Netzel Pub Tired Man Productions 2018

[107] Atlantis in the Caribbean by Andrew Collins Pub Bear & Co 2016

[108] The Secret of Atlantis by Otto Muck, Pub Collins 1978

[109] The History of the Discovery and Conquest of the Canary Isles by George Gas, Pub Elibron Classics 2006

[110] The UFO, ET, Alien Trilogy by Martin Thomas Self-Published 2025

[111] Bimini Road Anon self-published undated.

[112] Atlantis in the Caribbean by Andrew Collins Pub Bear & Co 2016

[113] Ibid

[114] The Secret of Atlantis by Otto Muck, Pub Collins 1978

[115] Atlantis the Andes Solution J M Allen, Windrush Press 1998

[116] The UFO, ET, Alien Trilogy by Martin Thomas Self-Published 2025

[117] Ibid

[118] Ibid

[119] Letters From Mesopotamia

[120] Bird-Headed Deity, Denver Art Museum web site

[121] Bronze Age Source of Tin Discovered The University of Chicago Chronicle 1994

[122] When the Fae Came by Martin Thomas. Self Published 2025.

[123] The UFO, E.T, Alien Trilogy by Martin Thomas. Self published 2026

Made in United States
North Haven, CT
11 April 2026

91261000R00082